ET's Are On The Moon and Mars

C. L. Turnage

Edited and Formatted by Bruce Stephen Holms

Timeless Voyager Press

ET's Are On The Moon And Mars

Entire contents © 2000 by Timeless Voyager Press

ISBN: 1-892264-01-3

Timeless Voyager Press
PO Box 6678
Santa Barbara, CA 93160

Front & Back Covers by Bruce Stephen Holms

Timeless Voyager Press

TABLE OF CONTENTS

Timeless Voyager Press

Timeless Voyager Press

PREFACE

Probes of Mars Reveal E.T. Ruins

The Mariner and Viking space probes, provided scientists with a vivid, yet startling look at the closest earth-like planet in our solar system, Mars. These probes were launched in an effort to examine the topography of the red planet, and thereby determine if any form of life exists there. The reaction however, of scientists studying the Mars data has been most peculiar. Puzzling if one considers the fact that concrete evidence of extraterrestrial life has indeed been discovered there. Not only in the Cydonian deserts of Mars, but in scattered ruins throughout the planet. Numerous mystery objects, that closely parallel archaic constructions of earth, have been captured on film by both probes.[1] Colossal pyramids, sphinxes with human looking faces, and remnants of what appear to be cities, all point toward the identity of their builders.[2] By this I mean, if one finds architecture quite similar to that found in the ancient world on earth, then there appears to be a connection between the two cultures. Since we know that the ancients were not themselves capable of space flight, it quickly becomes obvious that someone else was involved in this remote scenario. It became my quest, and that of other researchers, to discover exactly who was responsible for similar structures on both Mars and earth.

Timeless Voyager Press

NASA Remains Silent

One wonders as to why we haven't had an official word of these discoveries from NASA? Surely their scientists would have noticed the symmetry of some of the "formations," or the geometric layouts of these anomalies. Wouldn't they have noticed that some of the "rocks" are actually pyramidal? And, if they did notice, why haven't they alerted the rest of the scientific community? Are they afraid of what other scientists might discover? Or, is some other government faction outside of NASA pressuring them to be silent, and to cover up evidence of extraterrestrial life where ever, and when ever possible?

Extraterrestrials From Planet Beyond Pluto Visited Earth in Ancient Times

In 1976, the Viking Orbiter I successfully landed an unmanned robot craft on the surface of Mars. It remained in orbit a thousand miles above the planet. Its objective was to snap numerous unprecedented photographs from its bird's eye view of our enigmatic neighbor.[3]   That same year, the revolutionary book, The 12th Planet, by Zecharia Sitchin was released. It promoted a theory that extraterrestrial beings from a mysterious tenth planet in this solar system had visited earth in its remote past. Studying ancient Mesopotamian texts that predate the Bible, Sitchin discovered that the Sumero-Babylonian deities of the ancient world, were in fact beings from this tenth planet. (They numbered it as the twelfth planet, because they counted the sun and moon as planets). Its name appears in the seventh tablet of the Babylonian "Epic of Creation" tale. It is called NE.BI.RU, or "the planet of crossing."[4] The ancient symbol for the planet was the winged globe, usually with a cross in the center. The cross came to symbolize this body that periodically crosses between Jupiter and Mars during its perigee.[5] It was these extraterrestrials, the Nibirians, that influenced the Mesopotamians to build their stepped pyramids called ziggurats, and who later built the Egyptian pyramids and Sphinx.[6]

There is a mysterious celestial body known to modern scientists as well. They refer to it as Planet X, the tenth planet beyond Pluto. The X also stands for, previously "unknown" member of our solar system. Astronomers have detected its presence mathematically by the perturbations it causes in the orbits of the outer planets. NASA has issued the following two press releases regarding the possibility that a tenth planet does indeed exits: 87-32 & 87-33,

June 26, 1987. It is my belief, and that of Zecharia Sitchin, that this "tenth" planet, this Planet X, is in reality planet Nibiru. The planet where the Mesopotamian deities originated.

Alien Bases on Moon and Mars

In 1990, Sitchin published a fifth book about the Nibirians, Genesis Revisited. This book postulates that these "gods from space" established a base on Mars in ancient times. He also believes that they first came to earth nearly a half million years ago.[7] Intriguingly, this is the estimated age of the famous "face on Mars."[8] Bearing all this in mind, I decided to study the Mars data. I wanted to determine if there were additional clues that might help to establish if indeed these "gods from the heavens" were responsible for the curious constructions of Mars. As you will see, I believe that they did leave traces of their identities on Mars. This then led toward a desire to determine if these beings also had visited the earth's only natural satellite, the moon. To my surprise, they not only have in the past, but were actually present during the time our astronauts made their historic Lunar landings.

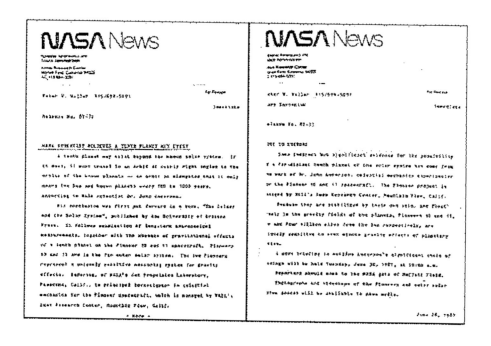

Timeless Voyager Press

ACKNOWLEDGMENTS

Most biblical verses are presented in their commonly translated form, being that in THE NEW AMERICAN STANDARD BIBLE. This Bible was employed because it was drawn from the most ancient texts available. It also provides one with the literal interpretation in side margins if the text has been altered. This results in greater precision of meaning. Newspaper articles, and various books and magazines have been cited. The drawings, and diagrams were executed by the author. The hardback version of the 12th Planet has been used for reference.

The author recognizes the Mariner 9 Principal Investigator, Dr. Arvydas J. Kliore, Mariner 10 Principle Investigator, Professor Bruce C. Murray, the Viking Orbiter Experiment Team Leader Dr. Michael H. Carr, and the National Space Science Data Center for providing the astounding Mars images published within this work. The author recognizes and thanks Dr. Michael H. Carr for the Apollo 11 and 12 moon data. The Johnson Space Center Mapping Science Laboratory, is recognized and thanked for their moon data. And, the Apollo 12 Experiment Team Leader, Mr. Frederick J. Doyle, Team Leader of Apollo 16, Dr. Richard J. Allenby, Lunar Orbiter III, IV, V, are thanked and acknowledged for the extraordinary Moon data. The author thanks NASA for their kind cooperation in helping her obtain the necessary Mars and moon data for this work.

Timeless Voyager Press

A special heartfelt thanks to Mark Jean of Film City, for his further processing of the Martian and moon enigma photos and other images for this publication. Many thanks to all those who contributed along the way, and to those specifically not mentioned, for without their successes, this volume could not have been written. The author would like to add that this work is not intended to denigrate any faith, or organization in any way. Nor is it meant to defame the name, or character of God. Its purpose is to expose the existence of an extraterrestrial civilization on both the moon and Mars. And, to identify those responsible for those constructions, as extraterrestrial beings from planet Nibiru, the planet from which the God of the Old Testament may have originated. The author remains open to other possibilities. She has come to these logical, however astounding conclusions, based on empirical evidence.

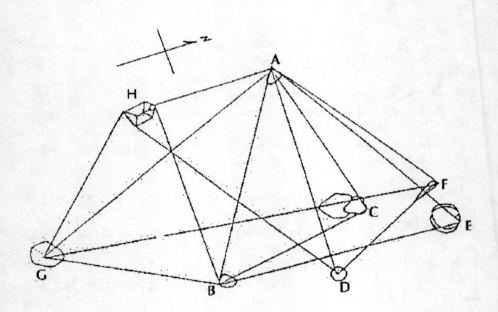

PART ONE

WHO WERE THE SUMERIAN
DIETIES OF ANTIQUITY?

Elohim Creators, Actually From Planet NE.BI.RU

After studying Mesopotamian texts, precursors to the Bible, and ancient Hebrew writings, Zecharia Sitchin, and other scholars, traced the origin of the Hebrew religion to the Sumerian and Babylonian deities of ancient Mesopotamia. He believed that they were also the Nephilim, the giants "who were cast down to the earth from heaven," of Genesis 6:1-4.[9] Sitchin also identifies these Nephilim, as the Elohim (plural deities) of the Bible. He then connects the Elohim, and the Nephilim to the Mesopotamian deities. By comparing the older Mesopotamian writings with the biblical texts, he and other scholars before him have shown how the first eleven chapters of Genesis were actually drawn from these earlier accounts. The Babylonian "Epic of Creation," or Enuma Elish, describes the creation of our solar system, and the entrance of planet Nibiru into this scenario.[10] The Assyrian tale "Atra-Hasis" preserves not only the story of the Great Flood, but the creation of man. It describes how the gods needed a slave to mine gold for them because of a rebellion of their own people stationed here on earth.[11] They decided to use the DNA of a primitive hominid already living here, and to fuse it with their own genetic material. Their experiments eventually proved successful, and mankind emerged on the face of the earth.

Timeless Voyager Press

Nibiru Is Biblical "Kingdom of the Heavens"

It was also Sitchin's assertion that these gods from heaven originated on the tenth planet beyond Pluto. The Sumerians called this the twelfth planet, because they counted the sun and earth's moon, as celestial bodies, or planets.[12] In both the Old and New Testaments, this planet is referred to as "the kingdom of the heavens." In the Hebrew writings, the Lord was said to come from olam, a Hebrew word meaning firmament or hard ground. In other words, the "kingdom of the Heavens" was to be found on a planet, and not floating about in some ethereal, or mystical realm. The Mesopotamian writings also described how "kingship was lowered from heaven."[13] In other words, Nibiru, or planet heaven, transplanted its own form of government here on earth. We know this because other writings revealed that Heaven was ruled by a king.[14] Even the Bible makes reference to the "king of Heaven," and insists that it is he who rules both heaven and earth. (See the book of Daniel).

Nibirians Travel To Earth Every 3,600 Years

In ancient times, the planet was symbolized by a winged-globe. It usually had a cross in the center. This enigmatic comet-like planet was called NE.BI.RU by the Sumerians, meaning as stated earlier, "the planet of crossing." The legendary planet is believed to cross between the orbits of Jupiter and Mars every 3,600 earth years, or one year in their reckoning of time. It is when Nibiru is in this particular conjunction that the Nibirians undertake trips to earth, stopping off first at Mars.[15] Curiously, the very type of structures which predominate their ancient Space Port of the Sinai, the pyramids and Sphinx of Egypt, are also the type of structures photographed by the Mariner, and Viking Probes on the shifting sands of the red planet.

An Egyptian step pyramid modeled after a Mesopotamian ziggurat.

The Pyramids of Egypt

The Bible Supports A Nibirian Origin for the Deity

Sitchin, and other scholars, believes that these Sumerian deities created man through genetic engineering. It appears that we are literally their descendents, and that we have their DNA.[16] In my first book, The Holy Bible is an Extraterrestrial Transmission, I demonstrated how the entire Bible, both the Old and New Testaments, support Sitchin's theory. And, I also determined that Christ is probably a hybrid offspring of the God of the Old Testament, Nannar-Sin, the Sumerian moon deity.[17] (See also my book: War In Heaven!). I also illustrated how the Bible reflects the last passing of Nibiru into our vicinity in the book of Exodus; while prophesies in Revelation, Matthew, and others, indicate the next and future passing.

In The Past Nibirians Provided Aid To Mankind

Sitchin points out that ancient cultures of the Middle East venerated the number 12. They were of the belief that twelve stood for the planet from which the Sumerian deities originated, Nibiru/Marduk. As if to bear up this hypothesis, archaic Middle Eastern deities were worshipped in groups of twelve.[18] The Bible carries on this Mesopotamian tradition with the symbolic use of the number twelve. (There were 12 tribes of Israel, and Christ had 12 disciples). Sitchin believes the evolution of man, was aided by the Nibirians. It appears that our advancement progressed through three distinct phases.[19] The Mesolithic phase began about 8,000 B.C., with the introduction of flint tools called microliths. The Neolithic phase of 7,000 B.C., lasted until people learned to smelt metals. It was during this period that man learned to grow crops, and

Timeless Voyager Press

domesticate animals. Finally, the emergence of Sumer, or early civilization, began around 3,000 B.C. It is marked by the invention of writing.[20]

Nibiru Last Passed Through The asteroid Belt in 1540 B.C.

If Nibiru reached perigee in 1540 B.C., then counting backwards in units of 3,600 (a year on Nibiru), it would have been present between Jupiter and Mars in 5,140 B.C. Next, it would have reappeared in 8,740 B.C. (the time of the great Flood recorded in the Bible). Counting backward another 3,600 years, it would have made an appearance in 12,340 B.C. The Mesolithic phase would have been in existence prior to the Flood. The Flood, could have been triggered by Nibiru's strong gravitational attraction. If there were an unstable polar ice shelf, its presence near the earth might have caused the ice to break free, and slide into the ocean, creating tidal waves, flooding, and tremendous storms. This most assuredly would have submerged most of the continents, destroying man's early achievements. Mankind would certainly have needed help from these advanced extraterrestrials to rebuild a devastated planet. So, from 8,740 to 5,140 B.C., man would have been taught the use of flints, domestication of animals, and crop cultivation. From 5140 B.C., civilization, cities and writing would have been the main focus, while the Hebrews became the people chosen to record and transmit the Holy scriptures around 1540 B.C., the time of their Exodus from Egypt. Moses received the computer coded Torah on Mt. Sinai, at some point, during this time period.[21] 2060 A.D. will usher in the New Age, the return of Christ and his planet.

1-A Kepler's Second Law

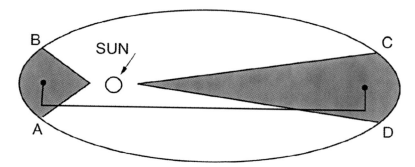

1-B Orbital cycle Of Nibiru/Marduk

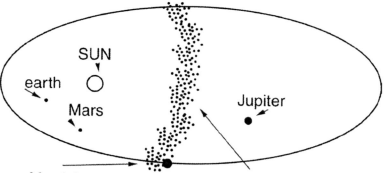

Marduk passing through asteroid belt 1540 B.C.

1-C "Long Day Of Joshua" 1440 B.C.

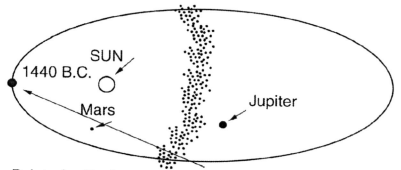

Point of critical passage as Marduk tries to break free of Sun

Timeless Voyager Press

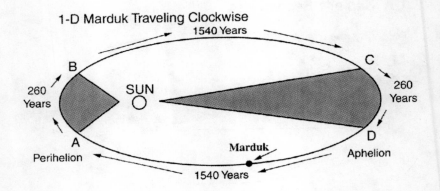

1-D Marduk Traveling Clockwise

1540 Years

B

C

260 Years

SUN

260 Years

A

D

Perihelion

Marduk

Aphelion

1540 Years

1-E Marduk moves slowly then quickly as it nears the Sun

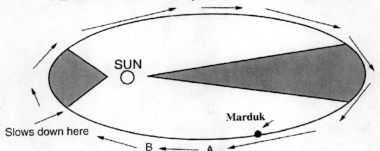

SUN

Marduk

Slows down here

B ← A

Approximately 50 Years to move from Point A to Point B
Birth pangs = A - B Portions of Orbit
Seven years of Great Tribulaition could begin any time prior
to 2060 A.D. as Marduk moves toward point B.

1-F Marduk moves solwly for about 3,435 years, then races toward the Sun during the last 165 years of its orbital cycle.

SUN

Marduk

100 years from
1540 - 1440
Perihelion

(2060 A.D.) 1540 B.C.

1,390 years from Aphelion to Pluto. 50 years from Pluto
to the asteroid belt. 100 years from asteroid belt to point
of critical passage. From its position beyond Pluto to " Place
of Crossing" = 65 years. 3,535 years of cycle completed
while 65 years remain.

PART TWO

TURN RIGHT AT MARS

Existence Of Extraterrestrial Life Is Impossible To Conceal

It has been impossible for the government to conceal the existence and activities of extraterrestrials since the modern UFO era began on June 24, 1947. That was when American pilot and businessman Kenneth Arnold sighted a formation of nine silvery disks flying over the Cascade mountains in the state of Washington. It was from Arnold's description of the UFO's that the term flying saucer originated.[22] Information regarding aliens is sometimes sprung on an unsuspecting public, purely by accident. For example, on July 25th 1976, the Viking Orbiter transmitted some rather startling pictures to earth. The most controversial of them is of a mile long human visage carved of stone, rising almost a half a mile above the surrounding Martian plateau located in Cydonia.[23]

The Man of Mars

This incredible "Face" on Mars, or Martian Sphinx as some have christened it, was first seen on NASA photo frame 035A72. The visage is of a man, who is clearly wearing some kind of headdress, looking remarkably like a space helmet. He has definitely detailed human features. His mouth is slightly open, and his eyes gaze straight up...if the observer happens to be out in space hovering above Mars that is![24] Ironically, Toby Owens, a member of the imaging team, and the original discoverer of the "Face" was said to have exclaimed,

Timeless Voyager Press

"Oh my God, look at this!" as he examined 035A72 as a possible landing site for the Viking Lander 2. Owen's unique discovery of the "Face on Mars," was immediately brushed off by the rest of the imaging team. They believed it to be a mere rock formation. Gerry Soffen, explained it to the press as a "peculiar trick of light and shadow."[25] The single most important discovery the world has ever known concerning the existence of extraterrestrial life, was momentarily dismissed. It was forgotten until three years later, when Vincent Di Pietro, an electrical engineer with fourteen years background in digital electronics and image processing came across it in a purported journal of extraterrestrial archaeology.[26]

At first, he to dismissed the "Man of Mars." Two and a half years later however, he came across the odd Martian visage among the archived NASA photos in the National Space Science Data Center, at the Goddard Spaceflight Center in Greenbelt, Maryland. He suddenly realized that it must be real, or as he put it, NASA wouldn't have put it boldly in the file." It was then that he turned to his friend and colleague, Gregory Molenaar, a computer scientist who shared a similar background. The two embarked on a bit of private research in order to improve the quality of the features in the NASA image by using a technique known as "computer enhancement."[27]

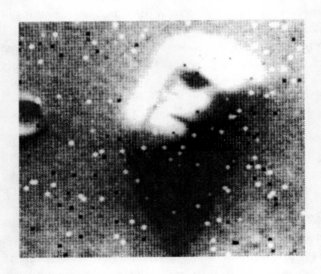

Viking Frame #035A72-"Monkey Face", Note "simian" appearance of face.

Computer Enhancements Provide Startling Details of Face on Mars

After trying conventional techniques of computer enhancement to reconstruct details of the eerie Martian visage, the two men became convinced that conventional procedures would not yield the desired results. It was then that they devised a brand new technique, somewhat different from the advanced mathematical manipulations of digital data that NASA has applied to spacecraft images. This new process was dubbed "Starburst Pixel Interleaving Technique," or "Spit Processing."[28] After applying this amazing new technique to 035A72, they found to their astonishment that the peculiar Martian object was essentially symmetrical. Half of the "Face" was in shadow, and where the other eye should be located was the suggestion of a deeper shadow. The brow, nose, and mouth also conformed to the normal geometry of a human-like countenance. Eventually, further computer enhancements revealed teeth in the mouth, and an eyeball in the socket![29] These findings began private research by dedicated scientists, who believed that despite the official prognosis, the Martian anomaly was worth checking into. This eventually led them to the subsequent discovery of scattered ruins in various regions throughout planet Mars.[30] (We will discuss these as the chapter progresses). Who would have ever imagined, that a mere 35 million miles from our own planet, separated only by the black chasm of space, lay the remains of a lost alien civilization, partially buried millennia ago by swirling Martian dusts?[31]

Mathematics of Face on Mars Points Toward Builders

This enigmatic "Face" also bears the mathematical signature of the its interplanetary builders. It is 1/360th of the planetary diameter of Mars, placing it within the sexagesimal system of the ancient Sumerians.[32] Zecharia Sitchin believes that these Sumerians inherited the sexagesimal system from their gods whom they called the Anunnaki, therefore it is arresting to find evidence of it on the red planet. He believes that the Anunnaki developed this system for their purposes, then scaled it down for human use.[33] The sexagesimal system is the only mathematical system deemed perfect for time reckoning, the celestial sciences, and geometry. In fact, the Sumerians had a term, DUB, that meant in astronomical lingo, the 360 degree circumference of the world, in relation of which they spoke of the curvature, or arc of the heavens.[34] 3,600, the supreme number of the Sumerians, and the number synonymous with the orbital period of Nibiru/Marduk, was signified by a 360-degree circle as well.[35]

Structures on Mars Resemble Ancient Structures of Earth

In addition to the puzzling Face, there are pyramids, and various structures clearly visible in other photos transmitted by the Viking and Mariner probes of Mars.[36] A striking collection of Martian artifacts has been dubbed "The City" by some researchers.[37] The main pyramid of the complex lines up parallel to the central midline of the Face, to within less than a degree.[38] Several other of its arresting collection of features, mathematically parallel those of the Face as well.[39] Some of the Martian structures appear to echo earthly structures found on the shores of Lake Titicaca in the Andes.[40] We must now ask the obvious question. Why would objects found on the plains and mesas of Cydonia, resemble structures found on earth? Did the builders of these Martian anomalies also erect ancient mystery monuments here on earth? If so, why?

Viking Frame #070-A11 (blow up) - Partial view of city complex.

Martian Topography Resembles Earth

Imagine for a moment that you have been transported through space to the rusted sands of the Cydonian desert. What would you see? Viking probes reveal topography not to different from that found in the vast expanse of the rusty deserts and craggy cliffs of the state of New Mexico, in the United States.

Ferric iron oxide responsible for the red sands of Mars, oxidized from the presence of oxygen reacting with iron, now colors the sands that would crunch beneath the soles of your boots.[41] A heavily eroded landscape, carved out by many centuries of wind and possibly water, would also confront you.[42] Looming hundreds of feet up from the desert floor, your eyes would quickly lock onto giant stone monuments of an exotic civilization not necessarily born on planet Mars...yet strangely familiar.

Viking Frame #070-A11 (entire frame). Comprehensive view of "City" and surrounding structures.

An Imaginary Survey of Cydonia Reveals Layout of Ruins

Standing at the base of the Face, you would see to your left, in the distance, massive pyramids. These are similar to their Egyptian counterparts, and are part of the City complex. (The City is located astride the North latitude 40.868 degrees, whose tangent equals e divided by pi, both universal mathematical constants).[43] To the right, an artificially designed Cliff, serves as a backdrop to the immense Face.[44] Below in the distance, your eyes would come to rest on another visage - one similar to the face of 035A72 that will be discussed later in the chapter. (I call this visage the "Fuzzy Face." See NASA

Timeless Voyager Press

photo frame number 070A13). To the lower right, a colossal construction appearing to be an even larger countenance would capture your attention.[45] Then, to the lower left, the peculiar five sided D & M pyramid staggers heavenward.[46]

Cydonian City and Pyramids May Have Been Part of Ancient Harbor

In the midst of what some call the "City Square," is the exact lateral center of "The City." Standing in that place a half a million years ago, one would have witnessed the rising of the Summer Solstice sun, immediately over the Face. In fact, the blazing orb of our star would have appeared to rise directly from the mouth of the god-like visage.[47] Who stood there and gazed at that marvelous spectacle eons ago? The Cydonian City appears to have been constructed around what once a harbor containing water in the planet's past.[48] Within the cluster of the pyramids called the City, there are seven major pyramidal forms, in various states of preservation. Among these are some half a dozen smaller objects. Some have pyramidal forms - but others have apparent "domes," "cones," "walls," and buried rectilinear markings to the southwest of the main "complex."[49] Bear in mind that the ruins of Mars may be partially buried in dust, just as the Egyptian Sphinx of earth once was.[50] Couple this with the fact that these ruins were photographed from 1,200 miles above the surface, and one quickly realizes that it is astounding that they can be viewed with the excellent resolution presented in the NASA images.[51] If it were not for the thin Martian atmosphere, then these monuments would probably blend right in with the topography of the planet.

The Face on Mars Has Counterpart on Earth

The layout of the haunting alien complex is rectangular, while the surrounding structures are aligned lengthwise, toward the rising sun. Similar layouts of terrestrial ruins attributed by Sitchin to the Elohim/Nephilim, are found in various places on earth. From ancient Mesopotamian ziggurats, Andean ruins, and early American pyramids, to the mysterious Nile Valley of Egypt.[52] Recently, a large visage resembling the "Face on Mars" was found in Marcahusi, Peru, carved into a rust colored mountain. The face is staring heavenward, just like its counterpart on Mars. It even seems to be sporting the same type of "space helmet!"[53] It would seem that these same, seemingly immortal space beings, constructed monuments not only on earth, but on Mars as well.

The Pyramids of Cydonia Similar to Those of Egypt

The main Cydonian pyramid is a colossal object, the well-preserved image of a classical pyramid. It is framed by two smaller objects, about the same size as the Great pyramid in at Gizeh, Egypt.[54] A very strange object, called the "Fortress," appears from some angles to be constructed in the manner of a Mexican pyramid. It has thick, straight walls, and an apparent interior space. This object has two walls forming a distinct right angle. What looks like a massive rocket lies next to the open end of the right angle walls.[55]

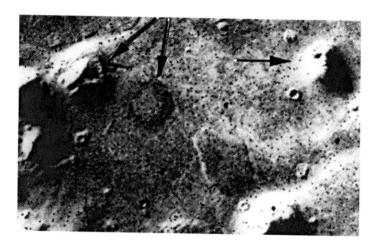

Viking Frame #76H593/17384 in relationship to "Face on Mars".

Cliff Serves As Backdrop to Face on Mars

Another peculiar formation dubbed the "Cliff" is perched on the edge of a massive impact crater. The absence of any concussion damage suggests that the 3.2 kilometer-long Cliff was constructed after the impact. Since this structure is located directly behind the Face (about 21 kilometers away), it probably served as a backdrop, against the rising of the Summer Soltice sun. To see the Face as a proper silhouette against the rising sun, it must be the only silhouette on the horizon. Because a crater lies directly behind the Cliff, it must have been necessary to construct an artificially flat horizon. This would have enabled the sun and earth to rise over the sharp edge of the planet. Thus, the construction of the Cliff would have been necessary to create this optical illusion on the horizon.[56] One of the most shocking and dramatic features of the Cliff, is that if it is viewed from the South, or directly overhead, it appears to be a stylized human face. It has eyes, and brow ridges, symmetrical cheekbones, and a mouth and chin, making it yet another "Face on Mars."[57]

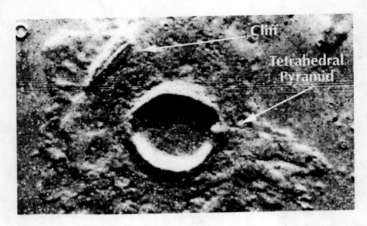

Viking Frame #035A74 (blow up) - Close up view of "Cliff" and Tetrahedral Pyramid". Notice how the eyes of the "Cliff" line up with the "Terahedral Pyramid".

Tholus Mound May Contain Stepped-Pyramid

The "Tholus" mound lies ten miles south of the Cliff. This is a strangely familiar formation, resembling a man-made earthen mound found in Great Britian, called "Silbury Hill."[58] Radio carbon analysis, dates Silbury Hill to around 2800-2600 B.C. Although it was excavated in 1968 and 1970, no central burial chamber was found, ruling out the possibility that it was created for burial purposes. It was discovered that beneath its earthen cover, there lies a masonry-faced structure of pyramidal form, with rounded corners, built on six levels. This peculiar British mound parallels the Egyptian step pyramid of the Pharoah Djoser. He ruled c.2630-2611 B.C. during the third Dynasty. This was about the same time as the creation of Silbury Hill! Curiously, this pyramid is in fact an energy node, sources of spiralling energy, of which Silbury Hill is by far the largest in the country.[59]

Viking Frame #035A74 (blow-up) - Close up view of "Cliff".
Note the stylized human face sporting eyes, elongated nose,
and small mouth (see drawing).

D & M Pyramid Appears to be Egyptian-Like

The five-sided D & M Pyramid, is a fantastic structure appearing to have the proportions of a human figure, with arms outstretched and legs spread. The man seems to be wearing an Egyptian style headdress on his head.[60] According to Sitchin, the family of Enki controlled Egypt in the remote past.[61] Legend has it that Ptah (Enki) created the land of Egypt in the beginning, by raising it up on dikes from the river Nile.[62] The Egyptian pyramids have been found to be simulations of Mesopotamian ziggurats under their skins of stone. Mesopotamian texts also provide evidence that the Sumerian deity Enki is responsible for their construction. Ancient texts as well as archaeological evidence attests to the close cultural, and economic links between Egypt and Sumeria.[63] Ironically, the five-pointed star, or pentagram, is the occult symbol of the biblical Satan. In Hebrew, the word Satan is "Sawtawn," and it means literally: Adversary.

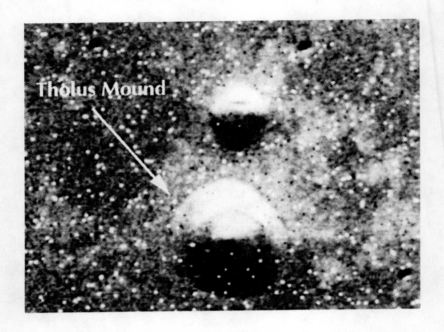

Viking Frame #35A74 - "Tholus Mound" of Cydonia, similar to
Silbury Hill, England; a step pyramis covered with earth.

Comprehensive view of "Famous Face", "Damaged Face",
"Fuzzy Face", "D&M Pyramid", "Tholus Mound", and Large Pyramids (far left).

This star-like building received its name from the two men who first discovered it, Vincent DiPietro, and Gregory Molenaar.[64] Its Pharonic head-dress points directly toward the famous Face.[65] Sitchin states that the ancient cultures of earth equated a star with a god, and with eternity.[66] (Particularly the cultures of ancient Mesopotamia, and those who emerged from them). The five-pointed star in Egyptian hieroglyphs, is the character seba used to denote "imperishable one." Every Egyptian temple served as a complex model of the whole cosmos, and the stars were intimately connected with the concept of immortality. The gods were the inhabitants of heaven, who resided among the stars.[67] The D & M pyramid is shaped in the exact form of the Egyptian hiero-glyphic seba, associated with immortality!

Viking Frame #070A11 (blow up) - Close up view of the "D&M" Pyramid, a five-sided structure with the figure of a man with outstretched arms and legs wearing a Pharonic style headress. (see drawing on next page.)

Face on Mars Identified

Since there appears to be an obvious connection with Egypt to the Martian City, it would be highly probable that the Face lying among the rusted dunes of Cydonia, gazing into eternity, is that of Enki, as further upcoming evidence indicates.[68] The anomalous visage affords us what I believe is tanta-mount to a photograph of "god." Curiously, when viewed from certain angles,

Timeless Voyager Press

the Face appears simian, or ape-like, and clearly resembles computer recon-structions of the primitive earth creature homo-erectus.[69] A half a million years ago, if indeed the Nibirians constructed the monuments of Mars as part of a Space Base, homo-erectus man beast would have roamed primeval earth in search of food.[70]

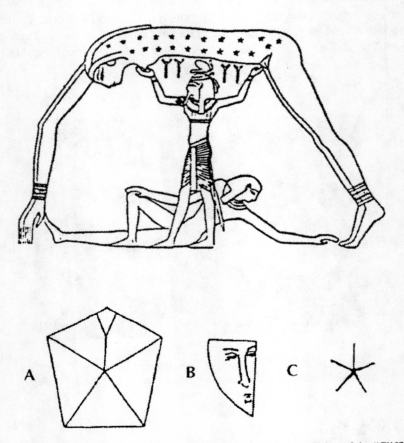

A.- Author's drawing of the "D&M Pyramid". B.- Author's conception of the "Cliff". C.- Author's drawing of Egyptian Hieroglyph, see "Reading Egyptian Art", by Richard H. Wilkerson, pg.131.

If Enki, the master scientist of the Elohim, created man from a mixture of homo-erectus and godly DNA, could then the monument known as the "Face" be his? Is it a stylized colossus, commemorating his ultimate scientific achievement - the creation of the human race? By combining his own features with the simian features of the homo-erectus, Enki may have provided man with a cleverly constructed clue to our interplanetary heritage, and to the identity of "god." Thus, he symbolically recreated his creation by using a double play on the theme of what he had accomplished. (The Sumerians must have inherited the "double meaning" of words, and symbolic themes, from their ancestors and creators...the Nibirians, for they also enjoyed using "plays on words. Even the Bible employs these technique). But, that is not all. The Face of Cydonia is more than a visage. There is some speculation that the side of the Face that is in shadow is feline. After taking frame 70A13, and subjecting it to a special form of computer image processing called "local contrast stretch," Dr. Mark J. Carlotto (TASC), was able to determine that the left side of the Martian Sphinx is what he believes to the be the face of the lion.[71]

The Cult Emblem of Enki in Utopia

The dark side of the Face might also be interpreted as reptilian (from my observation). This would fit in precisely with Enki's cult symbol - the Serpent.[72] Enki was the original settler of earth, and probably responsible for the establishment of the Martian base. It is logical that his cult symbol should be displayed in the various ways on the surface of the planet.[73] As you will soon learn, I have discovered his dubious Serpent token in the plains of Utopia. In frame 86A10, directly above another human-like head wearing a space helmet, located near the top of the frame, is the head of a viper! One eye is visible, scales, a slit of a mouth, and a nose. There is even a venom sack under the chin. It is surrounded by what appears to be debris. Careful scrutiny of the photo with a magnifying glass, reveals that the slit of the mouth is definite, the nose holes, the eyes, even the tiny scales! This is not such a surprising discovery if one considers the possibility that these Mesopotamian deities could be responsible for its creation, as well the other monuments of Mars. Is the biblical Serpent symbol of Satan (Enki), eyeing us from the rusted, wind blown ruins of a dead world, with the cold calculating eyes of a reptile?

Egypt And Mars Linked

A terrestrial link with Mars is undeniable. Right here on earth we have a similar representation of the Martian City. Cairo, the site of the two greatest pyramids, and the enigmatic Egyptian sphinx, was originally named "El-Kahira," from the ancient Arabic word El-Kahir meaning...Mars.[74]

Viking Frame #86A10 (blow up) - giant Serpent's Head and partial view of body among rocks. (See upper portion of photo). The cult symbol of the Sumerian Deity, Enki was the entwined Serpent. Notice: Eye, Nose, Mouth of Serpent Face. (See drawing below).

Egyptian texts ascribe communications functions to the famous Sphinx at Giza, and its face, according to inscriptions, is the likeness of Hor-em-Akhet, an epithet for Ra, the first born son of Enki who could soar to the farthest heavens in a celestial boat.[75] (Ra was known to the Babylonians as Marduk).[76] The Egyptian Sphinx was oriented so that its gaze was aligned precisely eastward along the thirtieth parallel. This was directly toward the Space Port in the Sinai Peninsula.[77] There are believed to be subterranean chambers beneath the Sphinx which enabled it to function as a communications center.[78]** If the purpose of the Egyptian sphinx was actually that of communication, then it is possible that perhaps the famous face on Mars was also used for that same purpose.[79] The Egyptian Sphinx is said to have received a "message from heaven," according to ancient texts.[80] If the Sphinx was part of the Space Port of the Sinai, could the message have been from Mars? Egyptian legend has it that the Sphinx could talk at one time.[81] Perhaps it could - by means of a radio transmitter, an amplifier, or loud speaking device.

Martian-Sphinx And Egyptian Sphinx Similar

At any rate, the discovery of an object on Mars that is similar to a terrestrial object such as the Sphinx, clears up many obscurities concerning the function of the Sphinx and the Martian Face. It also ties an ancient culture of earth to a civilization that existed on Mars at one time. Curiously, the Martian Sphinx is similar to the Egyptian Sphinx in the following ways:

1). Both have the visage of a man.
2). Both are wearing headdresses.
3). Both are sited N/S, E/W. on the e/pi latitude.[82]
4). Both are combinations of more than one creature. (This was called a chimera in ancient Mesopotamia. They were political emblems).
5). Both sphinxes are believed to have been used for communications.[83]
6). Both are placed in such a way that the left half is seen at the sunset of each respective planet.
7). Both are situated so that the right half is bathed in the rays of the rising sun.[84]
8). Both are extremely ancient

Comparison of Sphinx with Cydonian and Utopian Heads

Timeless Voyager Press

The pyramids and Sphinx of Egypt may have once been used to guide incoming spaceships along the landing corridor, whose apex was anchored on Mt. Ararat, the Near East's most visible natural feature. If this was the function of the face and pyramids of Cydonia, then one would expect a similar correlation between them and Olympus Mons, the largest Volcano in the solar system. It is a great source of geo-thermal power. And its a great natural point of reference that can be easily located from outer space, and used align landing spacecraft.[85]

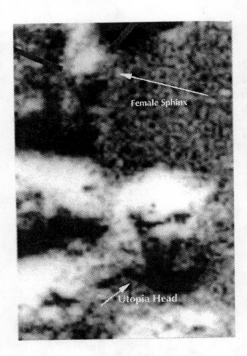

Another Face on Mars in Cydonia

From observations of NASA frame #070A13, yet another face exists that I have christened the "Fuzzy Face," due to its extremely blurred appearance. The obscure face has been missed by other investigators of the Cydonian City; and lies directly below the Famous Face at precisely 14 and a half centimeters on an 8x10 photographic reproduction of 070A13. It forms an isosceles triangle with both the Famous Face, and another anomaly that appears to be a larger, badly damaged face lying 10 and a half centimeters to the right of both the Famous Face and the Fuzzy Face. This large badly damaged face was first noticed by Maurice Chatelain, a retired NASA aerospace engineer, and reported in the Ancient Skies, a bi-monthly publications of the Ancient Astronaut Society.[86]

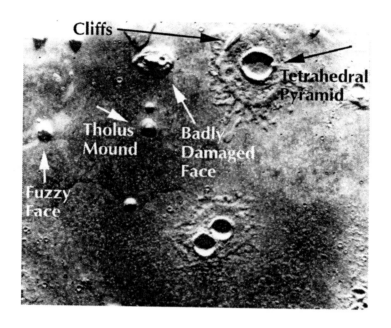

Viking Frame #035A74 - Complete view of "Badly Damaged Face", "Cliff", "Tetrahedral Pyramid", Fuzzy Face", and "Tholus Mound".

This twin, ghostly visage measures out at exactly the same dimensions as the Famous Face, making it one mile in length. It appears to be a male wearing a space helmet, and looking remarkably like the first, Famous Face. Two deep indentations are located where the eyes should be situated on a human face, and the dark slit of the mouth falls where it should naturally lie. Two small round "knobs" are located on the left side of the countenance. (These knobs are not found on the Famous Face). They seem to have been purposely placed there to make this anomaly line up with the triangular arrangement just described. When measuring the three formations, the triangular alignment does not work without the largest knob. (See drawing).

Some have probably dismissed this anomaly as a mere rock formation, just as the Famous Face initially was, however its position in the isosceles triangle described, indicates that it is an integral part of the Cydonian City layout. If the Famous Face is that of Enki as I speculated, could the large Badly Damaged Face be that of Anu? He was the best candidate for ruler of Nibiru, a half a million years ago. Could the Fuzzy Face represent the countenance of his other son, Enlil. Ancient texts state that Enki and Enlil were brothers. Enlil was the brother who was supposed to inherit the throne, and who sat at the right hand of power, next to his father. Enki sat at the left.[87] This configuration could portray the ruling triad of Nibiru as it was when the base was originally established.

If this is in fact the case, then the following scenario arises. The Sphinx of earth, believed to sport the face of Ra-Marduk, firstborn of Enki, would speak to the visage of Enki on Mars, (the Famous Face of Cydonia). The Cydonian Face might have in return spoken to a Sphinx of Anu on Nibiru, each time it passed between Jupiter and Mars during its perigee.[88] Thus, the first-born son of Enki would have spoken to his father, the firstborn son of Anu. (Though Enki was firstborn, he was not the legal heir, the source of many conflicts among the royal family).[89]

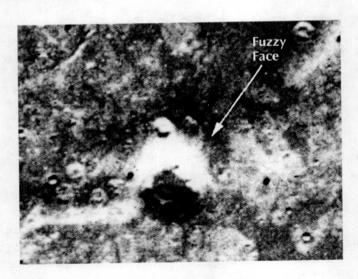

Viking Frame #070A13 (blow up) - "Fuzzy Face", making up part of the "Ruling Triad of Nibiru".

Layout May Represent The Ruling Triad of Nibiru

At a distance of 14 and a half centimeters from the Fuzzy Face, on an 8 x 10 NASA photo, and located to the right on frame #070A13, the construction known as the "tetrahedral pyramid" is found perched on the lip of the crater adjacent to the Cliff.[90] At a distance of thirteen centimeters to the extreme left of the Fuzzy Face, a very large pyramid looms in the distance. It is in perfect alignment with the Famous Face and the Badly Damaged Face. (See drawing, note precise distances between monuments, an obvious mathematical design). In frame #070A11, about 8 centimeters from the heart of the City Complex (that lies in the lower right center), there are found about a half a dozen small, cylindrical objects. The strange thing about these objects is that they display a uniform size, suggesting that they could possibly be something other than rocks. In fact, they closely resemble cigar-shaped objects photographed by NASA that appear to be flying over craters on the moon.[91] From their mathematical positioning, we can clearly see that the constructions of Mars are not simply natural formations, but are instead man-made structures conceived and established by someone of vast intelligence. Could that someone be the Elohim of the Bible, who are in reality the "gods of heaven and earth" described in so many ancient cultures?

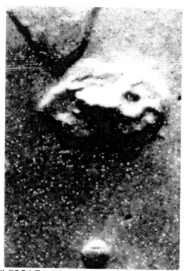

Viking Frame # 035A74 (blow up) - "Large Badly Damaged Face" discovered by Maurice Chatelain, probably part of the "Ruling Triad of Nibiru". (Possibly the face of the Nibirian ruler Anu).

Timeless Voyager Press

The Babylonian Cubit on Mars

Perhaps the most electrifying evidence that these ruins are the remains of a Space Base operated by the Nibirians on Mars, is the use of the Babylonian cubit used by both the Hebrew and Sumerian deities. A distance of one millimeter on the Viking photos is equivalent to 600 Babylonian cubits on the surface of Mars. This astute observation was made by the before mentioned Maurice Chatelain, and appeared in the same article in Ancient Skies that first published his discovery of the large, unfinished or Badly Damaged Face.[92]

Chatelain measured the distances and angles between the various monuments found within the City vicinity of the Cydonian region. He calculated that a millimeter on the Viking photos represented about 320 meters on the Martian surface. According to his observations and calculations, the polar circumference of Mars is about 21,333,312 meters, compared to the value of earth at 39,999,960 meters. This is a ratio of 8/15 of one meter, or 533.333 mm. If a distance of one millimeter in the Viking photos indeed represents 600 cubits on the planet Mars, it is reasonable to assume that the Babylonian cubit originated on Mars. And, that it was derived from the polar circumference of Mars, then imported to earth by the Mesopotamian deities millennia ago. Consequently, planetary dimensions measured on earth in meters are the same as those measured on Mars in Babylonian cubits.[93]

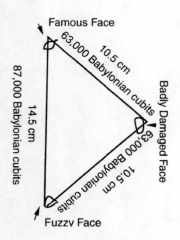

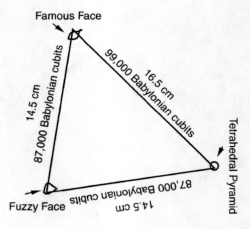

This is a strange coincidence indeed, considering that God order Noah to build the ark in cubits. Ezekiel saw a temple on earth measured in cubits. Even the Hebrew Ark of the Covenant, and some of the contents of the Ark room (the Most Holy Place), were designed using cubits. Whose measuring system were ancient humans using? Chatelain also calculated the height of the D & M pyramid to be 1800 cubits (3,600 divided by 2 = 1800). That means it is six and a half times the height of the Great Pyramid of Gizeh![94] This could mean that the builders of the monuments of Mars, are also the originators of the Babylonian cubit.

Man Is Not Descended From Martians

Man himself could not be responsible for these enigmatic structures, for a half a million years ago, homo-erectus was barely eking out an existence in a primitive animal-like fashion. And, it is unlikely that man is descended from "Martians" either. Even if Martians had existed millennia ago, and escaped the death of their world by colonizing earth, there is geological evidence for the devastation of earth in periodic 3,600 year cycles (which some have attributed to the passing of Nibiru) in support of the "gods from Nibiru" hypothesis. There is the insistence by the Sumerians and Babylonians that their gods did indeed originate on Nibiru, not Mars. According to ancient texts, the red planet was merely a stepping stone between Nibiru and earth, as the heavenly planet traveled between Jupiter and Mars during its annual circuit. It was the site of a Nibirian base.

The Nibirians Turn Right At Mars

Biblical prophesy indicates that it will again be in this position in the end times, when Christ returns. Thus Martian ruins and artifacts, seem to fit in theologically with Judeo-Christian teachings, and pre-biblical Mesopotamian texts. Sumerian planetary lists describe the planets as the Nibirians passed by them on their voyage to earth. This was called the "zone of flight." Intriguingly, Mars was "MUL.APIN," or "planet where the right course is set." A circular Mesopotamian tablet has been found, showing a route map for the journey from Nibiru to earth undertaken by Enlil, which graphically reveals that upon approaching Mars, the Elohim turned right. In the Akitu festivals of Babylon, Mars was termed "the travelers ship." This led Sitchin to suggest that it was here that the gods stopped off at their Martian base for cargo transfers, or perhaps to resupply before proceeding onward toward Nibiru.[95]

Timeless Voyager Press

An ancient Sumerian cylinder seal depicts an astronaut out by Mars, represented by a six-pointed star. He waves to a winged member of an earth based ancient astronaut, standing beneath the symbol for earth, seven balls in the upper left-hand corner. Between the two worlds, suspended in outer space, is what appears to be a spacecraft. Sitchin suggests that the two astronauts might have been telling each other that a spaceship had left Mars, and was on its way to earth.[96] Because the visages of Mars are clearly human, they probably depict the gods of the Babylonian cubit - the biblical Elohim who created man in their "own image" according to Genesis.

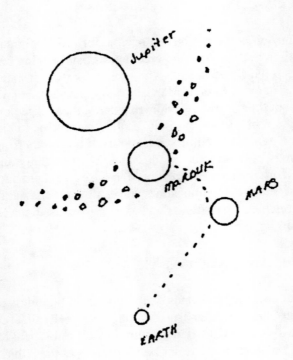

Cubit Used in Heaven

Interestingly, the New Jerusalem "space city" spoken of in Revelation is measured in Babylonian cubits! Could this be the rescue vessel, destined to airlift the "Elect" from earth, before civilization is destroyed by the coming global deluge? This was prophesied by Christ when he warned that the coming of the son of man would be just as it was in the days of Noah. Will this rescue vessel called the New Jerusalem, deliver them to Heaven after first stopping off at Mars? "And the city is laid out as a square, and its length is as great as its width; and he measured the city with the rod, twelve thousand stadia (15 hundred miles); its length and width and height are equal. And he measured its wall, one hundred forty four cubits (seventy-two yards) according to human measurements, which are also angelic measurements." (Revelation 21:16-17).[97]

Here the Bible specifies that the Babylonian cubit used by humans is actually the same measuring unit used in heaven! Is this why Babylonian cubits are found on Mars? Will they also be discovered to be the measuring unit of the moon as well? Revelation 21:10-11, John describes the New Jerusalem, saying that it has the "glory" of God. Does this terminology mean it is a similar vessel, perhaps with a glow like that of heavenly vessels described by Ezekiel?

Yet Another Face on Mars

Besides the features found in the Cydonia region of Mars, bizarre structures were discovered in the Utopia, Elysium, and Deuteronilius areas. In Utopia, a mottled "Head" displays similarities to the Famous Face. Both have an indentation above the right eye, and each has on a helmet, or headdress.[98] (See comparison photo on back cover). In Sumerian times, the planet Mars was associated with the Mesopotamian god Nergal.

Viking Frame #86A10 - "Utopian Head" and surrounding structures.

Timeless Voyager Press

Another name for Mars was "Simug," meaning "smith." Nergal was the chief god in charge of mining, and there is distinct evidence of mining in Utopia, where this particular head was discovered to lie. Could this head, that almost seems to be laughing at us from the red sands of Mars, be the likeness of Nergal, one of Enki's sons? Large structures surround the head, and to the left of it, one can clearly see what appears to be roads. In the lower right hand corner of the picture, what some believe to be the profile of a face peeks out.[99]

An Incan Face on Mars

The most phenomenal head in Utopia, is one that appears to be Incan, and that dwarfs the above described head. This huge head is a bas-relief, a flat, three-dimensional image of a man, meant to be viewed only from above. The visage is in profile, looking straight ahead. The mottled "Head" appears to be emerging from the mouth of the larger profile. (The mouth forms the dark opening to the left, and slightly above the head, while the white area to the left appears to be a lower lip). The long, almost hook-like nose is directly above the mottled "Head." A large, dark, gash is evident above the bridge of the nose, and appears to represent an eye. The curve of the forehead is obvious above the nose and eye. At the lower left of the bottom lip, a false beard, like that worn by Pharoah, seems to be connected to the chin.[100]

A Female Sphinx on Mars

The surrounding terrain of 86A10 contains several other anomalies supporting the hypothesis that these structures were artificially designed. There is an almost square structure to the extreme lower left, with a curved wall resembling an ark, or bow. A serpentine like road, conduit, or cable, snakes its way across the center of the photo. It is connected to what appear to be two structures in its middle. Yet another feature can be discerned with careful scrutiny in frame #86A10. Directly behind and above the left side of the head, there appears to be an Egyptian style Sphinx. Careful examination of the photo with a magnifying glass, reveals that this oddity is almost identical to the one in Gizeh Egypt. It appears to have a feminine human face, and a large base. The head supports an Egyptian style headdress similar to that of the Egyptian Sphinx. Note the striations below the chin, in the chest and shoulder area. These same striations appear on the Great Sphinx of Egypt. This anomaly appears to have been weathered, either by wind, or by water, just like the Great Sphinx in Egypt.[101]

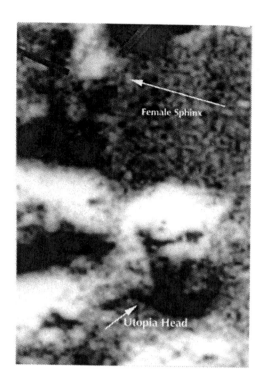

Viking Frame 86A10 Utopia terrain surrounding faces. Note inter-
secting lines which may be conduits or roads. Also, small "domes"
or pyramid-like structures in upper left.

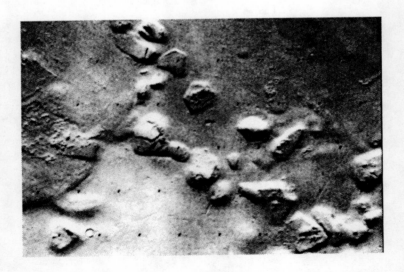

Viking Frame #086A07 - Five sided Structure. Near center of photo.
Five sided structures are not "normal" natural features on any known planets.

Pentagonal-Shaped Structure on Mars

In frame 86A07, an odd pentagonal-shaped structure lies among what appears to be large scattered ruins.[102] There are also other unusual surface features that warrant further investigations. The first appears on the Mariner frame 4209-75 (not pictured in special photo section). It exhibits strange indentations, with radial arms protruding from a central hub. The area in which the unusual feature was observed is located at 1.9 degrees latitude, and 186.4 degrees longitude.[103] According to this technical explanation, this formation is caused by the melting and collapse of permafrost layers, indicating the current presence of water on Mars.[104]

Incan City on Mars

An area dubbed "Inca City" by the Mariner-9 team, is found in frame 4212-15. It is located at -80 degrees latitude, and 64 degrees longitude, not far from the Martian south pole. This "City" consists of a collection of pyramidal features, and other "unusual formations." The official NASA description of this region is this: "Striking geometric markings, resembling the ruins of an ancient metropolis."[105] A most astounding feature on the Martian terrain is what has been described as a gigantic "runway." It lies at 34.7 degrees latitude and 212.8 degrees longitude (Cebrenia Quadrangle MC-7). This runway is found in the Utopia region, the same area of the head, and profile of photo #86A10. It appears that this runway lines up east to west, and is built on the upslope, toward the volcano Hecates Tholus, roughly centered 200km to the South East. Its exact position is North West on the lower slopes of the volcano. The runway is 0.9 kilometers wide, and is a tilted slab structure 5 kilometers long.[106] The tremendous stone slab is quite similar to a huge platform of unknown origin discovered in Baalbek Lebanon, which Sitchin suggests may be a former runway of the Elohim-Nephilim.[107]

A Martian Landing Platform

The south side of the runway is occupied by a series of sheared blocks appearing to be smaller structures. The western end of the runway is covered with erosional debris from the mountain above. There may be a parallel fault on the north side of the runway, making it a true graben, however, this is entirely covered with debris and rubble. The "bumps" on the mesa suggest large buildings.[108] (NASA Frames #086A and 86A-08) This runway lies halfway around the planet from the Cydonian artifacts. Perhaps the specifically aligned structure, with regular knobs along its length, might be some kind of accelerator, used to launch spacecraft from the planet's surface. (Proposals to build similar structures on the moon for launching payloads impractical with current rocket technology, have been seriously suggested for decades.[109]

Viking Frame 86A10 - Large style base relief of possible face (see outline).
Note roads or irrigation channels and conceivable surrounding structures.

The runway latitude - 35 degrees north of the equator - is the same as the maximum obliquity of Mars, during its million year "nodding cycles." This would strategically place the orbit of the inner moon Phobos, in the same plane. This suggests the possibility that it may be used in some connection with the launching of space shuttles, perhaps in much the same way our astronauts would use the moon as a "jumping off point" on a trip to Mars.[110] Some unusual tracks have been discovered on Phobos, as well as some of the other moons of the solar system.[111] The elevation of this runway structure above the mean datum, almost 4 km, is consistent with another main requirement of an accelerator: that on Mars, (unlike the moon) it must be built as high as possible, to avoid the dragging effect of the thin Martian atmosphere.[112] Could this be the very runway used by the Nibirians to launch heavy payloads, before sending them to earth, or on to Nibiru, as implied by ancient texts?

Nazca-Like Markings on Mars

Frame 651A06 striations forming an arrow that appears to have been etched into the Martian surface by something. This marking reminds one of the Nazca markings in Peru. Ancient Astronaut Researchers attribute the Nazca markings to the ancient gods. Could the arrow be pointing the way to something on the Martian surface that is visible from the air? Could it be a landing platform? It is intriguing that the cuneiform symbol for "arrow" and "life" is the same for "ti."[113]

A Huge Circular Wall

Another area that seems to contain a large circular wall with a door, or gap is found in frame #070A11. I call this the "amphitheater." In the close-up view of this feature, one can clearly see a large raised object in the center, at almost the same angle as the gap. It appears as though someone or something blasted into this "wall," and knocked some of it inside the structure.

Timeless Voyager Press

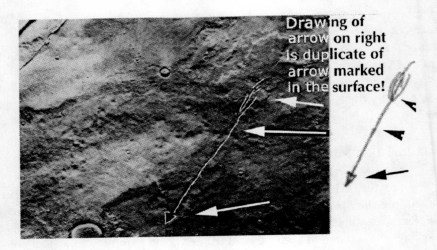

Viking Frame #651-A06 (blow up) - Gigantic arrow etched into Martian surface reminiscent of Nazca markings in Peru.

More Pyramids in Elysium

Additional anomalies have been photographed by the Mariner 9 space probe in the Elysium region. Frame #4205-78, and 4296-23, reveal scattered pyramids throughout the area.[114]

Timeless Voyager Press

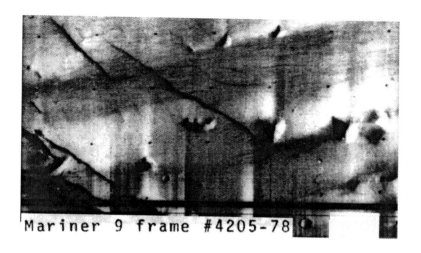

Mariner 9 frame #4205-78

"Pyramids of Elysium"

Evidence of Catastrophe on Mars

After closely examining the ruins, and archaeological remnants of the lost Martian civilization, we come to a seemingly fundamental truth - there were at one time, intelligent life forms on the red planet. But, what happened to this once thriving civilization? Where have those that once populated it gone? What happened to Mars? Why has it become a seemingly dead world? There are strong indications that it was not always so. There is still surface water on Mars. Viking 2 discovered an area where frost was present on the ground. It is known that the Martian North Pole consists of water ice. Scientists of the former Soviet Union believe that water may still flow beneath the surface of dried out river beds on Mars.[115] Photographic evidence provided by Viking 1 and 2 probes, supports the theory of large amounts of water on the surface of Mars in the past. Chryse Planitis indicates flooding by large quantities of water, while Vallis Marineris once released running water that carved large channels into the rocky terrain of Mars. It would also appear that Mars still has water vapor in its atmosphere. There are cloud formations. There cannot be clouds without some water, and some atmosphere. Some past catastrophe appears to have altered Mars, ripping away its atmospheric mantle, and permitting most of its surface water to sublimate into space.[116] What cataclysm destroyed the once earth-like planet only a few million miles from home? And,

Timeless Voyager Press

what does the destruction of Mars have to do with man? Is it a frightening omen of things to come?

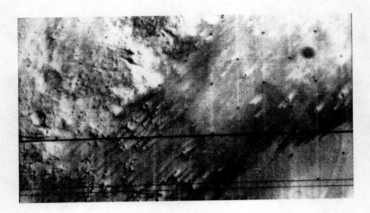

Mariner 9 Frame #4205-075 Cloud shadows on Mars

Nuclear War Waged on Mars!

The New Testament book of Revelation speaks of a war fought in outer space. "And there was War in Heaven (planet Nibiru), Michael and his angels waging war with the dragon. (Michael led the military forces of the home planet). And the dragon and his angels waged war..." The "dragon and his angels" were the Adversarial forces spoken of throughout the Bible. I call them the serpent faction, because they use for their cult emblem, or political symbol, the serpent or dragon. Revelation also states that the dragon swept away a third of the inhabitants of heaven, and threw them to earth. On third of the people of Nibiru were forced into exile, and had to come to earth. However, as Nibiru attained apogee, and the military threat from the home planet was eliminated, they most likely returned also to occupy their Mars, and moon bases.

The frightening nuclear war that most likely spread to these bases, as the Nibirians of the home planet fought to push the renegades off their world. The logical thing for the Serpent faction to do would be to head for the Mars base and launch an attack from there. Richard Hoagland in his book the Monuments of Mars, makes an excellent case for nuclear war on Mars. He points out that the Face on Mars is damaged, and that the D & M pyramid, and the Honeycomb complex, also appear to have been damaged by some type of force.[117] There are also an unusually high number of heavy impact craters, dotting many areas of Mars. These may have been caused by sophisticated nuclear weapons.[118] Mars is scarred with craters laid out in geometric patterns, like those used by bombers in the Vietnam conflict. This geometric dropping pattern is used to insure the greatest possible destruction of human life in a region. On the moon, traces of radiation are still being emitted from Tycho and Copernicus. One can see the white rays emanating from the crater Tycho with the naked eye. Whatever caused that crater blew up before it hit the planet's surface, and not on impact.[119]

The Strange Fate of Phobos 2

On March 28, 1989, Soviet Mission Control acknowledged that the probe Phobos 2, failed to communicate with earth as scheduled after completing an operation around the Martian moon Phobos. Stable radio contact was never again made with the probe. The next day, a high ranking official, Nikoli A. Simyonov, of the Soviet Space agency Glavkosmos made this shocking state

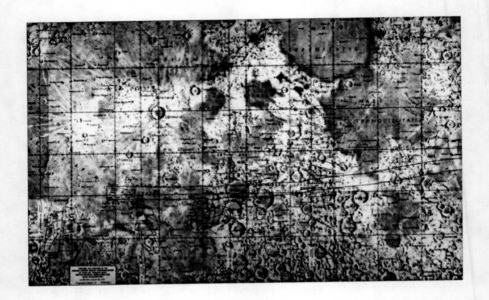

Note white rays of the craters," Tycho" and "Copernicus". (NASA microfiche)

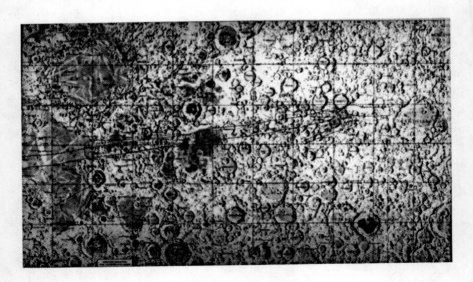

Note strange white markings.

ment: "Phobos 2 is ninety-nine percent lost for good."[120] The following day the Spanish daily La Epocha published this puzzling article: "PHOBOS 2 CAPTURED STRANGE PHOTOS OF MARS BEFORE LOSING CONTACT WITH ITS BASE." The article explained that the Phobos 2 was orbiting above Mars when Soviet scientists lost contact with it, however it did manage to photograph an unidentified object on the surface of Mars, seconds before losing contact.[121]

The T.V. newscast Vremya devoted a long segment of its broadcast concerning the Phobos 2, to a couple of very important pictures of a large shadow, and to the final and most tantalizing picture of an inexplicable "thin ellipse" that can be clearly seen. According to researchers from the Soviet Union, the object is some 20 kilometers (12.5 miles long). A few days earlier, the probe had recorded an identical phenomenon 26 to 30 kilometers (16 to 19 miles long).[122] (Bear in mind that the Nibirians are described as being giants. This might explain why their craft are so extraordinarily long). The anomaly seen in the Phobos 2 transmission was a thin ellipse with very sharp marquise points, and the edges stood out sharply against a kind of a halo on the Martian surface. It has been described as something that is between the spacecraft and Mars, because the Martian surface can be seen below it. The object was seen by both the optical and infrared (heat seeking) camera. As the last picture was halfway through, the Soviets saw something that should not have been there. Some have speculated that the Soviets saw something crash into the Phobos 2, abruptly interrupting the transmission.[123]

The Soviets were slow to release this last frame. However, Zecharia Sitchin, a man endowed with what some might call "gifted insight," conjectured that the Phobos 2 (which was spinning either from an impact with an unknown object, or some computer malfunction) was shot down by aliens for intruding upon their Martian base.[124] In 1994, as if to confirm his eerie speculation, an American T.V. program, Sightings broadcast this final frame. There was what appeared to be a long, white, cylindrical object with what appeared to be a window in the front! This peculiar rocket-like object, looked remarkably like white cylindrical objects photographed by American astronauts on moon missions.[125] (There is a recently published Russian photo of the white cylindrical object. This image can be seen in the book Extraterrestrial Archaeology, by David Hatcher Childress, Adventures Unlimited).

Phobos 2 Tried To Photograph "Sensitive" Martian Site

Curiously, the probe had been photographing an area of straight lines in the area of the Martian equator; some of the lines were short, some longer, some thin, some wide enough to look like rectangular shapes "embossed" on the Martian surface. Arranged in parallel rows, the pattern covered an area of some six hundred kilometers (230 square miles). Oddly enough, the Phobos 2 had used its infrared (heat-seeking) camera to photograph this perfect geometric pattern.[126] Could it be that the Phobos 2 was destroyed over Mars for unintentionally photographing a "top secret" area in use at this time by aliens?

Mars Observer May Have Met Similar Fate

Replica of Russian Phobos 2 Probe "destroyed" during Mars Mission!

In 1992 a Titan rocket launched NASA's billion dollar Mars Observer from Cape Canaveral, Florida into earth orbit. Another rocket then kicked the Observer on toward its ultimate destination - the red planet. The probe was designed to study Martian terrain, climate and weather using sophisticated instruments. It was also armed with a camera that could spot an object the size of a Volkswagen Beetle from a 234-mile-high orbit.[127] On Saturday August 21st, the Mars Observer failed to respond to efforts made to contact it. A newspaper article published by the Ft. Worth Star Telegram on August 25, entitled "Observer Hangs Heavy On Scientists," (Associated Press) revealed that scientists did not know whether it had flown past Mars, was destroyed over Mars, or was disabled. NASA ultimately blamed a computer malfunction for their inability to contact the probe.

Ironically, the same paper carried another article on August 25 entitled, 'Spaced Out Plot? NASA accused of keeping Martian City, Life a Secret' (Associated Press). Included in the article was a photo of the Face on Mars. The article detailed efforts by the Mars Mission, (a group founded by Richard C. Hoagland to study the Martian artifacts) to suggest that NASA may have purposely disabled the Mars Observer so that it could not take pictures of the artificial structures of Cydonia. Hoagland believes that a rogue group within NASA intentionally shut off the spacecraft so that it could not make an orbital survey of the City. He has been trying for 10 years to get NASA to investigate the Face on Mars; and he alleges that scientists for the agency have dismissed his research and that of his colleagues in his organization.

Hoagland believes that a study conducted 30 years ago suggests a "McCarthy-esque fear of fundamentalists and religious fanaticism" within NASA. Hoagland believes that the space agency is reluctant to admit that there is evidence that intelligent life once existed on Mars. This suppression of the truth because of a fear of religious people, has caused NASA to ridicule, and trivialize Hoagland's efforts to study this phenomenon. And, he believes that this is why a rogue group within NASA "pulled the plug" on the Mars Observer. NASA of course, denies this accusation. Hoagland wrote an article published in UFO magazine titled: "Did NASA Disobey the Prime Directive?" In the article he suggests that the Mars Observer Mission has temporarily converted into a "stealthy Cydonia Mission." He believes that during the secret mission, NASA will be acquiring, secret high-resolution images of the Face, pyramids and artificial structures of Cydonia. After a specified period of data gathering, NASA will secretly resurrect the Mars Observer. Then the images

retrieved secretly, will be altered by means of sophisticated computer enhancements and released to the public as live Cydonia television. This, he believes
will be an attempt by NASA to trick the public into believing that there is no
basis for the claim that artificial structures exist on Mars.[128]

However, there is always the possibility that the Mars Observer was
actually destroyed over Mars. Did the Nibirians view the probe as an intruder
on their base, attempting to take it out of the sky? Did it manage to transmit
any photos before its destruction? What really became of the Mars Observer
remains to be seen...

*The sexagesimal mathematical system is based upon the number 60.
**A recent video production titled, "The mystery of the Sphinx," uncovers the
findings of a researcher that the Sphinx is about 11,000 years old, and that there
is a hidden subterannean chamber beneath it.
Measurements in centimeters refer to 8 x 10 original NASA photos of described
anomalies, according to Maurice Chatelain. See drawing of distances between
monuments for estimates in Babylonian cubits on Mars surface.

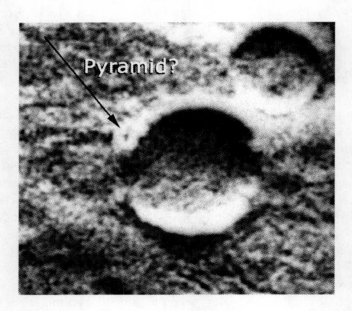

Viking Frame #043A02 - Twin Craters in Deutronilius. Note how something has
removed sections of crater walls in lower left crater. Notice what also appears
to be a pyramid on left wall of lower right flank, possible indications of mining.

Viking Frame #043A02 - Furrowed ground in Deuteronilius. Is someone harvesting crops on the surface of Mars?

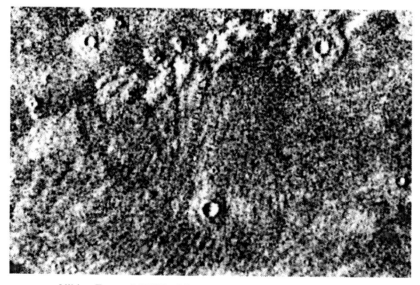

Viking Frame # 17599 - More furrowed ground in Deutronilius.

PART THREE

THE EAGLE HAS LANDED?

American Astronauts Photographed Extraterrestrial Ruins on the Moon

What did the Apollo Astronauts really find on the surface of the moon? In the winter of 1974 issue of UFO Report, an analysis was made of astronaut voice tapes. These tapes revealed the startling observations of the perplexed spacemen upon seeing what appeared to them to be unusual, and unnatural formations, resembling the ruins of a dead civilization.[129] Down through the ages, legends conveyed stories of the cities that existed on the moon in earth's remote past.[130] Oddly, astronomy magazines in the U.S. have published pictures of pyramids, domes, bridges and even crosses on the sterile lunar landscape. According to the famous writer Charles Fort, European astronomers claimed in scientific publications, to have actually seen such ruins during the last century.[131] The ancient book of Enoch, reports that he was taken to outer space by "two very big men" as "never seen on earth." Their "faces shone like the sun," and their "eyes were burning candles." These extraterrestrial giants showed him the "power of the moon's light."[132] (Compare this description of the "giant men," of Enoch, to the "man" Daniel saw along the Tigris river in Babylon, and it seems that they are the same type of extraterrestrial. Daniel 10:4-6). Were these curious "men" Nibirians, and did Enoch have the unique experience of being the first earthling to see a Nibirian moon base?

Emblems Carved Out of Lunar Rock, Point Toward Ancient Mesopotamia

According to the April 1954 issue of Harvard Universities Sky and Telescope magazine, a bridge exists and has been photographed on a lunar mountain range. John O'Neill, Pulitzer Prize winner and science editor of the New York Herald Tribune, along with two British astronomers, Patrick Moore, and H.P. Wilkins, confirmed this story. The bridge is said to span part of a mountain range adjoining Mare Crisium. If Mare Crisium is the dried out basin of an actual lunar ocean, then this bridge would be comparable to the Golden Gate Bridge in San Francisco, which also spans two mountains next to an ocean.[133] A most startling photo, taken through a telescope, portrays a series of platforms spanning 30 miles in an area known as Archimedes. There are five of these platforms that appear to be giant letters, or symbols carved out of rock, obviously meant to be seen from space. One of these is clearly a bird resembling an eagle.[134] This is rather ironic, since the Mesopotamian symbol for pilot was that of an eagle. These angelic pilots were often referred to and depicted as eagle men.[135] Was it a mere coincidence that our astronauts stated: "the eagle has landed" when they arrived at the moon? Pyramids, domes, tunnels and what appear to be water reservoirs with reinforced walls, have been observed on the lunar surface as well.[136]

Pyramids, Human-Like Head, and Other Anomalies on Moon

Very compelling photos of anomalous objects that resemble sculptures depicting human heads, similar to those photographed on Mars, have also been photographed on the moon. Located inside a giant crater, this bizarre collection of oddities resides to the left of the crater edge. To the extreme left, near a small crater, there appears to me to be a human head turned to the left. The head is in profile, and an eye, nose, mouth, and hair can be made out. What appears to be a gigantic pyramidal object looms to the right. What could be an upside down head is found in the upper center of the photo as well.[137]

UFO's Photographed Over Moon

Eyewitness reports by the astronauts themselves, and photographic evidence made by NASA, reveal cigar-shaped objects in flight over the planet, and at rest on the ground.[138] In fact, a strange collection of gigantic, cylindrical objects can be seen resting on the floor of the crater Archimedes. These anomalies seem to resemble the flying object photographed by the Phobos 2 space probe shortly before it was "destroyed" over Mars. These oddities are at least twenty miles long, and three miles wide![139] Glowing disk-shaped objects were photographed in the lunar skies, and in space near the moon.[140] Unknown objects have sometimes been observed emerging from craters.[141] One glowing UFO was photographed, as it hovered above an Apollo astronaut while he worked on the lunar surface.[142] The light from the object was so bright it lit up the astronaut, his equipment, and the surrounding terrain with an eerie bright light.

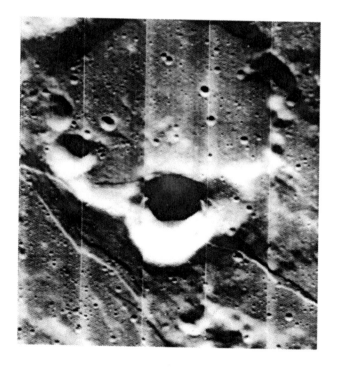

Lunar Orbiter IV took Frame #LOIV no.161-1 of the Damoisseau D
Notice two double wall-like structures on the crater to the right and left
of inside. Are the white lines flowing NW to SE, through the crater riverbed?
In lower left corner there appear to be small pyramids.

Timeless Voyager Press

Apollo 11 Frame #AS11-44-6609 is #AS11-44-6611 from another angle. Note large head in upper center of photo. It appears tp have toppled from its base and rolled over backward. Note eyes and mouth on the upside down visage.

Hovering glowing object casts shadows towards camera!

Apollo Frame #AS-12-497319, a glowing object hovers over Apollo 12 astronaut in this NASA photo. Notice the bright light reflecting off the astronaut and the equipment. Even the lunar surface is illuminated from above!

Astronauts Voice Transmission Reveals Evidence of Extraterrestrial Civilization on the Moon

During the Apollo 16 Mission (April 16-27, 1972), Charles Duke, Thomas Mattingly, and John Young landed in the Descartes highlands. In a conversation between the astronauts on the moon, and Mission Control,

Duke remarked, "These devices are unbelievable," and that he wasn't "taking a gnomon up there."

Young responded with "O.K., but man, that's going to be a steep bridge to climb."

Timeless Voyager Press

What devices had they seen? And what bridge?

Duke continued with "Tony, the blocks in Buster are covered - the bottom is covered with blocks, five meters across. Besides, the blocks seem to be in a preferred orientation, northeast to southwest. They go all the way up the wall on those two sides, and on the other side, you can barely see the outcropping at about 5 percent. Ninety percent of the bottom is covered with blocks that are 50 centimeters and larger."[143]

What blocks did Duke refer to? A meter is equivalent to 39.37 inches. Therefore the blocks described as five meters across, were over 16 feet across. They were in a "preferred" orientation of northeast to southwest. Preferred by whom? The beings that placed them there? Was this a huge landing slab like the one at Baalbek Lebanon? Or, was it something else entirely?

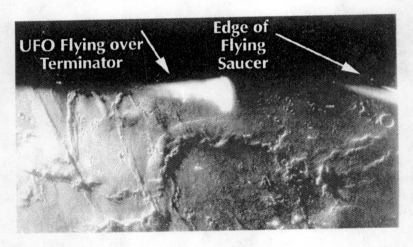

Apollo 11 Frame # 11-37538 photo of glowing cigar shaped object close to moon.
Observe part of saucer shaped object in upper right corner near terminator.

Huge, Crystalline Structure Discovered

Duke found something else peculiar at the Descartes site, "Right out there...the blue one I described from the Lunar module window is colored, because it is glass coated. But, underneath the glass it is crystalline...the same texture as the Genesis rock...dead on my mark."

Timeless Voyager Press

What was Duke describing here? Something blue, coated in glass, but crystalline underneath - with a texture like that of the Genesis rock. It is impossible to ascertain what this was, but whatever it might have been, it was large enough to be seen from the window of the lunar module.

Young : "Mark, it's open."

Duke: "I can't believe it!"

Young: "And I put that beauty in dry!"

What was this large, blue, glass-crystalline structure?

Duke then proceeds to remark, "These EMU's and PLSSs are really fantastic."[144]

It is obvious that the astronauts were speaking in code, disguising what they were actually referring to. This must have been something the NASA scientists believed might be found on the moon. Otherwise, they would not have created code-words for the astronauts to use upon finding such objects. And judging from the excitement in their remarks, it seems unlikely they were finding pretty rocks.

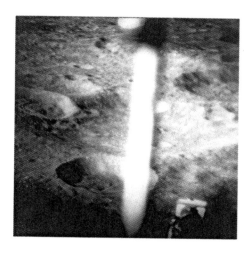

Apollo 16 Frame #16-192338 Hasslebald photo of cigar shaped aerial object flying over the lunar surface. Identical to object photographed by Phobos before disabled.

Astronauts Speak of Domes, Structures, and Tunnels

At one point, after remarking on the beauty of the moon's scenery,

Duke: "You'd have to be there to see this to believe it - those domes are incredible!"

Mission Control: "O.K., could you take a look at that smokey area there, and see what you see on the face?"

Duke: "Beyond the domes, the structure goes almost into the ravine that I described. And, one goes to the top. In the northeast wall of the ravine you can't see the delineation (layout). To the northeast there are tunnels, to the north they are dipping east about 30 degrees."[145]

Obviously, Duke had been referring to some type of structure, one that went almost into the ravine. Nearby are tunnels. Did these tunnels lead inside the moon? Apparently, Apollo 17 astronauts Eugene Cernan, Ronald Evans, and Harrison Schmidt who landed in the Taurus-Littrow Valley (December 7-19, 1972), discovered more domes, and geometric formations.

Astronaut Evans to Mission Control: "...I guess the big thing I want to report from the back side is that I took another look at the - the - clover leaf in Aitken with the binocs. And that southern dome (garble) to the east."

Mission Control then asked if there were any differences in the color of the dome and the Mare Aitken there?[146]

Condorcet Hotel on the Moon

Evans: "Yes, there is...that Condor, Condorsey, or Condorcet, or whatever you want to call it there. Condorcet Hotel is the one that has got the diamond-shaped fill under in the uh-floor."

Mission Control: "Understood. Condorcet Hotel."

Evans : "Condor, Condorcet, Alpha. They've either caught a landslide on it, or its got a - and it doesn't look like (garble), in the other side of the wall in the northwest side."

Again the astronaut uses code. And, who is the mysterious "they've" who have caught a landslide?

Mission Control : "O.K., we copy that Northwest wall of Condorcet A."

Notice they stated "wall." A wall of a structure referred to as Condorcet Hotel, which has a diamond shaped fill in the "floor."

Evans : "The area is oval, or elliptical in shape. Of course, the ellipse is toward the top."[147]

What does this refer to that has a "top"?

Astronauts: "Barbara"

Another very strange conversation took place with the Apollo 16 astronauts, again in code.

Capcom to Young: "Just a question for you John. When you got halfway, we understood you lopped around south, is that right?"

Young: "That is affirm. We came upon — Barbara."

What does this refer to? Barbara is a female name meaning "beautiful barbarian." Had they come across a sculpture of some alien goddess? Or, an even more intriguing possibility, a living inhabitant of the moon base occupied by the Nibirians? Joseph F. Goodavage made an appointment with NASA geologist Farouk El Baz at the National Aeronautics and Space Museum. He asked him what "Barbara" stood for. El Baz stated that it was an enigma, and probably a code word.[148] A code word for what?

Bright Spots, Halos, and the Number One

The following Apollo 17 conversation implies that the astronauts were not alone on the moon.

DMP (lunar module pilot): "What are you learning?"
Capcom: "Hot spots on the moon, Jack?"

DMP: "Where are your big anomalies? Can you summarize them quickly?"

Capcom: "We'll get that for you quickly on the next pass."

CMP (command module pilot): "Hey, I can see a bright spot down on the landing site where they might have blown off some of that halo stuff."

Capcom: "Roger. Interesting. Very - go to KILO, KILO."

CMP: "Hey, its gray now, and the number one extends."

Capcom: "Roger. We got it. And we copy that its all on the way down there. Go to KILO. KILO on that."

CMP: "Mode is going to HM. Recorder if off. Lose a little communication there huh? Okay, there's bravo. Bravo, select. OMNI. Hey, you know you'll never believe it. I'm right on the edge of Orientale. I just looked down and saw the light flash again."

What are "hot spots" on the moon? And what "big anomalies" are there? What was the "halo stuff" that blew off? What was the mysterious light that flashed in Orientale? The unusual dialogue continues.

Capcom: "Roger. Understand."

CMP: "Right at the end of the rille."

Capcom: "Any chances of ...?"

CMP: "That's on the east of Orientale."

Capcom: "You don't suppose it could be Vostok?"[149]

(Vostok were Russian probes that never reached the moon, they only orbited the earth). Apparently Vostok was a code word used to describe non-American craft, perhaps even extraterrestrial craft. Blocks, beaches, and terraces on the moon

While on the moon, some of our Apollo 16 astronauts may have seen indications of alien handiwork, such as strange constructions, or disturbances.

Orion: "Orion has landed. I can't see how fat the (garble)...this is a blocked field we're in from the south ray - tremendous difference in the albedo. I got the feeling that these rocks may have come from somewhere else. Everywhere we saw the ground, which is about the whole sunlit side, you had the same delineation the Apollo 15 photography showed on Hadley, Delta and Radley Mountains...."

What is going on here? How could rocks on a supposedly uninhabited world come from somewhere else? What is this "blocked field they keep referring to? Whatever it was, it was markedly different from the rest of the moon's topography. Enough so, that the astronauts wondered about it themselves.

Capcom: "O.K., go ahead."

Orion: "I'm looking out here at Stone Mountain and its got - it looks like somebody has been out there plowing across the side of it. The beaches - the beaches look like one sort of terrace right after another, right up the side. They sort of follow the contour of it right around."

Capcom: "Any differences in the terraces?"

Orion: "No, Tony. Not that I could tell from here. These terraces could be raised out of (garble) or something like that..."

Casper: (Mattingly in lunar orbit overhead): "Another strange sight over here. It looks - a flashing light - I think its Annbell. Another crater here looks as though it is flooded, except that this same material seems to run up on the outside. You can see a definite patch of this stuff that's run down inside. And, that material lays, or has been structured on top of it, but it lays on top of things that are outside and higher. It's a very strange operation."[150]

Timeless Voyager Press

Now what in the devil was that plowing? Who, or what has terraced part of the moon? What is this "stuff" that ran down inside the crater? What are the things that are outside and higher? I definitely agree with them that it seems to be a "very strange operation."

The Moon Rings Like a Hollow, Metallic Ball

A very strange thing about the moon is that it rings like a hollow sphere when a large   object hits it. During the Apollo missions, ascent stages of lunar modules, and spent third stages of rockets crashed on the hard surface of the moon. Each time these caused the moon according to NASA, to "ring like a gong, or a bell." On one Apollo 12 flight, reverberations lasted from one hour to four hours before ceasing.

During an Apollo 17 flight, astronauts commented to one another about this very phenomenon.

LMP: "Was there any indication on the seismometers on impact, about the time I saw a bright flash on the surface?

CAPCOM: Stand by. We'll check on that, Jack."

LMP: A UFO perhaps, don't worry about it. I thought somebody was looking at it. It could have been one of the other flashes of light."

What other flashes of light is he talking about? Notice how casually and matter-of-factly they conjectured the presence of a UFO? Apparently, the real possibility that they might encounter one already existed in their minds before they got to the moon.

CAPCOM: "We copy the time and..."

LMP: "I have the place marked."

CAPCOM: "Pass it on to the back room.

LMP: O.K. I've marked it on the map, too."

CAPCOM: "Jack, just some words from the back room for you. There may have been an impact at the time you called, but the moon is still ringing from the impact of the S-IVB impact. So it would mask any other impact. So they may be able to strip it out another time, but right now they don't see anything at the time you called."[151]

LMP: "Just my luck. Just looking at the southern edge of Grimaldi, Bob, and-that Graben is pre-Mare. Pre-Mare!"

CAPCOM: "O.K., I copy on that, Jack. And as long as we're talking about Grimaldi we'd like to have you brief Ron exactly on the location of that flashing light you saw...We'll probably ask him to take a picture of it. Maybe during one of his solo periods."[152]

Flashing Lights, Watermarks, and Another UFO

Notice that the Capcom repeats that it was a "flashing light." It was therefore no meteor impact that they were witnessing. Again, the lunar command pilot specifically mentions the word UFO. Maps were marked, and photos taken at the sites of these occurrences. During this same mission, "watermarks" were discovered.

LMP: "Al Baruni has got variations on its floor. Variations in the lights, and its albedo. It looks almost like a pattern, as if the water were flowing up on a beach. Not in great areas, but in small areas around the southern side. And, the part that looks like the water-washing pattern is much lighter albedo. Although I cannot see any real source of it. The texture however, looks the same."

CAPCOM: "America, Houston. We'd like you to hold off switching to OMNI Charlie until we can cue you on that."[153]

While the astronaut were discussing the "watermarks" the sighting of the UFO occurred. The conversations then returned to the watermarks. (NASA has recently confirmed ice on the moon. This indicates that at one time, there may even have been liquid surface water, if not now).

Astronauts See "Constructions" on the Moon

DMP: "O.K. 96:03. Now we're getting some clear - looks pretty clear - high watermarks on this."

CMP: "There's high watermarks all over the place there."

LMP: "On the north part of Tranquilitatis. That's Miraldi there, isn't it? Are you sure we're 13 miles up?"

CAPCOM: You're fourteen to be exact Ron."

LMP: "I tell you there is some Mare, ride or scarps that are very, very sinuous - just passing one. They not only cross the low planar areas, but go right up the side of a crater in one place, and a hill in another. It looks very much like a constructional ridge - a mare-like ridge that is clearly as constructional as I would want to see it."[154]

Was the astronaut seeing a bridge? Was it constructed on the edge of the crater, connecting it to a hill? What did the fellow mean when he said "as constructional as I would want to see it?" Was this a coded way of saying he was seeing what he wanted, or expected to see there - evidence that intelligent beings had built themselves a bridge?

Strange "Tracks" on the Moon

Apollo 15 astronauts David Scott, Alfred Worden, and James Irwin went to the Appenine mountians of the moon on July 26 - August 7, 1971. There they discovered strange "tracks."

Scott: "Arrowhead really runs east to west."

Mission Control: "Roger, we copy."

Irwin: "Tracks here as we go down slope."

MC: Just follow the tracks, huh?"

Irwin: "Right. We're (garble). We know that's a fairly good run. "We're bearing 320, hitting range for 413...I can't get over those lineations, that layering on Mt.Hadley." I can't either. That's really spectacular."

IRWIN: "They sure look beautiful."

Scott: "Talk about organization!"

IRWIN: "That's the most organized structure I have ever seen!"

Scott: "It's (garble) so uniform in width."

IRWIN: "Nothing we've seen before this has shown such uniform thickness from the top of the tracks to the bottom."[155]

Oddly, Fred Steckling, author of "We Discovered Alien Base on the Moon!" published a NASA photo of a large object leaving behind definite "stitchmarks" by some form of belted vehicle. (No.67-H-1135). He also published a nice drawing of what the vehicle may look like. He believed that this object might have been a mining vehicle, or soil testing device.[156] Harrison Schmidt, was a trained geologist, the only civilian ever to walk on the moon, he also reported seeing tracks.

Schmitt: "I see tracks running right up the wall of the crater."

MISSION CONTROL: (Gene Cernan) "Your photo-path runs directly between Pierce and Pease. Pierce Brava, go to Bravo, Whiskey, Whiskey, Romeo."[157]

Secret Channel Used For Transmission When UFO's Appeared

Apparently, whenever something was discovered, the astronauts and CAPCOM switched to a prearranged code, meaningless to the public. They sometimes even switched to an alternate publicly non-monitorable channel. Our men on the moon frequently encountered UFO's.

CAPCOM: "You talked about something mysterious..."

ORION: "O.K., Gordy, when we pitched around, when we pitched around, I'd like to tell you about something we saw around the LEM (Lunar Excursion Module). When we were coming about 30 to 40 feet out, there were a lot of objects - white things - flying by. It looked as if they were being propelled or ejected, but I'm not convinced of that.

CAPCOM: "We copy that Charlie."[158]

Were these white "things" that flew by UFO's? That they were being propelled or ejected implies that some intelligence was maneuvering them.

Astronauts Not Alone on Sea of Tranquility

Oddly, the very first Apollo Mission to land on the moon on July 20, 1969, encountered spaceships. Apollo 11 astronauts Neil Armstrong, Michael Collin and Edwin Aldrin were surprised to find that they were not alone when they descended in the lunar module on the Sea of Tranquility at 4:17 P:M. Both Armstrong and Aldrin saw UFO's shortly after the historic touchdown. One of the astronauts referred to a light in, or on a crater during the television transmission, followed by a request from mission control for further information. Nothing more was heard, publicly that is.

Frightened Astronauts Saw Huge UFO's

Back on earth, ham radio operators that had their own VHF receiving facilities that bypassed NASA's broadcasting outlets received the transmission that NASA briefly interrupted. (Live television broadcast became disrupted for 2 minutes due to a supposed overheated camera). Otto Binder, a member of the NASA space team, describes how when the two moon walkers, Aldrin and Armstrong where making their rounds some distance from the LEM, Armstrong clutched Aldrin's arm excitedly and exclaimed:

ARMSTRONG: "What was it? What the hell was it? That's all I want to know!"

MISSION CONTROL: "What's there?...malfunction (garble)...Mission Control calling Apollo 11..."

APOLLO 11: "These babies were huge, sir! ...Enormous! ...Oh, God! You wouldn't believe it! ...I'm telling you there are other spacecraft out there...lined up on the far side of the crater edge! They're on the moon watching us!"[158]

How could a television camera on the most sophisticated endeavor of the twentieth century break down for 2 minutes? Could an overheated camera cool off enough after only two minutes to begin functioning again? The T.V. camera was one of the most essential elements of the entire project. Christopher Craft, director of the base in Houston, commented that NASA censored three things that day — image and sound, and that there were other spaceships on the moon.

Here is what was actually said that historic day.

ARMSTRONG & ALDRIN: "Those are giant things. No, no, no - this is not an optical illusion. No one is going to believe this!"

HOUSTON (Christopher Craft): "What...what...what? What the hell is happening? What's wrong with you?"

ARMSTRONG & ALDRIN: "They're here under the surface."

HOUSTON: "What's there? (muffled noise) Emission interrupted, interference control calling Apollo 11.

ARMSTRONG & ALDRIN: "I say there were other spaceships. They've lined up on the other side of the crater!"

HOUSTON: "Repeat, repeat!"

ARMSTRONG & ALDRIN: "Let us sound this orbita...in 625 to 5...Automatic relay connected...My hands are shaking so badly I can't do anything. Film it? God, if these damn cameras have picked up anything - what then?"

HOUSTON: "Have you picked up anything?"

Timeless Voyager Press

ARMSTRONG & ALDRIN: "I didn't have any film at hand. Three shots of the saucers or whatever they were that were ruining the film."

HOUSTON: "Control, control here. Are you on the way? What is the uproar with the UFO's over?

ARMSTRONG & ALDRIN: "They've landed here. There they are, and they are watching us."

HOUSTON: "The mirrors, the mirrors - have you set them up?"

ARMSTRONG & ALDRIN: "Yes, they're in the right place. But, whoever made those spaceships surely can come tomorrow and re-move them. Over and out."[159]

The astronauts actually stated that someone, or something was under the moon's surface. In my book, War In Heaven!, I pointed out that the exiled Adversarial faction of the Nibirians were exiled to earth. Revelations chapter twelve describes how they have been thrown out of "heaven" and can't go back. It says that they are on earth, under the earth, and under the sea. Perhaps they are also on their old bases on the moon and Mars.

Forget the North Pole - Santa Claus is on the Moon

Walter Schirra, aboard Mercury 8 was the first astronaut to employ the code name "Santa Claus" to indicate the presence of the UFO"s beside space capsules. (The astronauts were forbidden to speak directly of UFO's, or any-thing unusual they had seen in outer space). His announcements were barely noticed by the general public. However, some people began to sit up and take notice when James Lovell on board the Apollo 8 command module came out from behind the moon and said for everyone to hear, "We have been informed that Santa Claus does exist! Even though this transpired on Christmas day 1968, many people sensed a hidden meaning to those words that was not diffi-cult to decipher.[160] What, or who had Lovell seen on the dark side of the moon that confirmed his belief in "Santa Claus?"

Moon Believed to Have Vegetation, and Water

Oddly enough, Apollo 8 took color photographs of the dark side of the moon that revealed vegetation green in color, that goes through seasonal changes.[161] (Let us bear in mind that a study of the moon's phases reveals that the moon seems to change shape, because we see different parts of its sunlit surface as it orbits the earth. Like the earth, half the moon is always lighted by the sun's rays, except during eclipses. Sometimes the far side of the moon, is in full sunlight, even though it is out of our view).[162] This means that the far-side of the moon does get sunlight. Though the moon is a somewhat hostile environment temperature wise, that does not mean that it would be impossible for vegetation to exist. After all, there are places here on earth such as Antartica, that support vegetation, though it would seem unlikely.

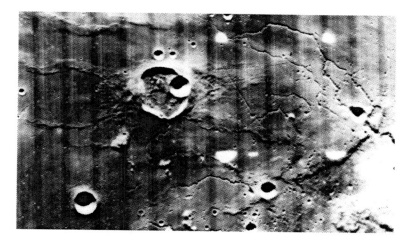

Lunar Orbiter Frame #151-3 photo of "Krieger" area. To the right of "Krieger" observe what might be five white triangular "pond-like" constructions reflecting sunlight. Fred Steckling says that these are faint clouds reflecting sunlight in lower right corner of photo.

What some believe to be water on the lunar surface has been photographed in what appear to be man-made reservoirs, and naturally occurring lakes and ponds. These anomalies are located on the dark side.[163] The lunar soil appeared to be the same color as cocoa. It looked almost wet, the astronauts reported. In fact, photographs that reveal the astronauts footprints in the lunar soil, show them as being very well-defined. Almost like footprints in wet sand. This seems to be evidence of a lot of water, trapped in the lunar soil. If

Timeless Voyager Press

the moon is really just a dry, desert-like environment, then why would the footprints be well-defined? Wouldn't they appear like footprints in a desert? The Italian scientist Dr. Maria, studied the moon's rocks and soil. It was his conclusion that if all the water, now existing on the moon, was released to the surface, a shallow ocean would entirely cover it.[164] The existence of water, thin atmosphere, and possible vegetation, could also produce clouds. Clouds are actually the product of condensed moisture, rising to a density altitude, from the warm ground below. For clouds to exist, moisture on the ground must exist. The densest part of the atmosphere is found in the valleys and craters of the moon, close to the lunar "sea-level." This same law applies to earth. Logically it should apply to the moon. Lunar clouds seem to hug the mountainsides, much like monsoon clouds over tropical islands on earth. In fact, NASA photos exist of these clouds on the moon.[165]

Tsiolkosky Crater taken by NASA Lunar Orbiter V Frame #MR 121-2. Note small island in center of crater, surrounded by what some believe to be shallow water. Back side of Moon.

Ironically, if the dark side does in fact harbor vegetation and water, it would be the logical choice for the installation of a hidden lunar base, safe from the prying eyes of mankind. Now, if there is water, then this means that in all probability, there is a thin atmosphere. If there is vegetation, even scant, there should be some release of oxygen. There should also be carbon dioxide, since plants must take in carbon dioxide in order to produce oxygen. As if to support the existence of a thin atmosphere, two minutes and twenty seconds before touchdown on the lunar surface, they experienced some drift. The spacecraft

Apollo 16 Frame #16-19-228 of King Crate area. Note similarity between these "Tholus Mounds" and the ones on both earth and Mars.

actually drifted along at a seventy-five foot altitude, and had to be landed manually. (The astronauts were in a bind fuel-wise. The drift would not have been due to thrusting. They had to conserve every precious drop of fuel they had, in order to have enough for the return trip to earth).[166]

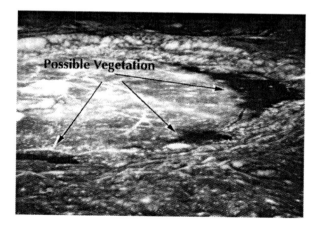

Apollo 12 Frame #12-7419 photo of the crater Humboldt area of the Moon. Notice the dark patches that look very much like thick vegetation. The entire area appears to be covered with growth. Could the white lines be irrigation channels, roads, or even Nazca-like markings?

Timeless Voyager Press

Gemini and Apollo Astronauts Saw UFO's

On November twelfth, 1966, James Lovell, and Edwin Aldrin, in Gemini 12 saw two UFO's at slightly over a half a mile from the capsule. These were photographed repeatedly. This also happened to Frank Borman and James Lovell in Apollo 8 on Christmas Eve 1968, as well as to Thomas Stafford and John Young aboard Apollo 10 on May 22nd, 1969. The UFO's showed up during the orbit around the moon, and on the homeward flight of Apollo10.[167]

Astronauts Subjected To mind Altering Force in Space

During the flights, the astronauts frequently felt as if mysterious external forces were trying to take over their minds. Some of the astronauts had psychological problems and changes in their personality after their missions in space.[168] Were the Nibirians probing the minds and observing the activities of their human creations? Is this the real reason why the American and even the Soviet Space Programs have never tried to establish a base on the moon, the next logical step if man is ever to reach Mars, and eventually even the stars?

Apollo 13, A Disaster Caused By A UFO?

There was talk that the Apollo 13 Mission carried on board a nuclear device that could be set off to make measurements of the moon's infrastructure. The detonations would show on the charts of several recording seismographs, placed in different locations. According to rumors, the unexplained explosion of an oxygen tank in the service module of Apollo 13, on its ill-fated flight to the moon, was actually caused by a UFO. It is believed that this object was

Timeless Voyager Press

following the capsule to prevent the detonation of the atomic charge, that could possibly have destroyed, or compromised some extraterrestrial moon base.[169]

Former NASA Director Warns That We Are Being Watched by Aliens

Perhaps man would be wise to heed the words of Albert M. Chop, deputy public relations director of NASA (circa 1965), as quoted for True Magazine, January 1965:

"I've been convinced for a long time that flying saucers are interplanetary. We are being watched by beings from outer space."[170]

Timeless Voyager Press

PART FOUR

AUTHOR'S CONCLUSION

The Builder's of Pyramids on Mars Connected to Egyptians

The presence of pyramids, an Egyptian style sphinx, and human-look-ing heads, on the red sands of Mars, leads one to believe that the builders of such monuments were probably connected to the Egyptians in some way. The Egyptian civilization had its roots in the Sumerian culture. Therefore it seemed

Timeless Voyager Press

likely that evidence of Mesopotamia would be found there as well. I soon discovered exactly what I was looking for - the cult symbol of a Sumerian Deity - the giant Serpent's head of Utopia. Further support that the Sumerian deities may have once been on Mars was discovered mathematically. The distances and layout of these monuments on Mars can be measured in precise Babylonian cubits. For similar architecture to also be discovered on the earth's moon is further evidence to support their presence there as well. As you can see from NASA photos presented in this work, these god-like extraterrestrials have not entirely disappeared from the scene. Which brings us to the question of why they are on the moon and Mars now.

They Never Die

As Sitchin pointed out in his writings, these extraterrestrials from planet NE.BI.RU, are very long living. They seem almost to be eternal beings. In fact, the word for Nibiru in Farsi, the native tongue of Persia, or Iran, is "Nimiru." It means: "They never die." Such incredible life spans make them appear to be "god-like" to mere human beings. We must also remember that time is different for them. One of their orbits around the sun is equivalent to 3,600 earth years. Jesus Christ has been gone from earth for about 2,000 earth years, or maybe seven or eight "months" in their time. When he left, according to various verses in the New Testament, there was a "War in Heaven," that was left unfinished. (See the book of Revelation, chapter twelve). The "army of Heaven," or the forces of Christ are prophesied to return in the "end times," and destroy the renegade Serpent faction. All of this will take place in less than one year of their time, although a couple of thousand years have taken place here on earth. And, this could explain why there are bases on the moon and Mars still being manned by these extraterrestrials. The Serpent faction, exiled from planet Heaven, will undoubtedly employ their bases on these two planets to launch counter-attacks against Nibiru as it attains proximity to earth. (I provide details of the interplanetary struggle in "War In Heaven!").

Crop Circles and Agriglyphs Are Likely Messages from Serpent Faction

As the crop circles and agriglyphs appearing in the grain fields of the world seem to indicate, the Nibirians are soon to make an appearance here on earth. Unfortunately, these "transmissions" in the grain, are most likely messages from the Serpent Faction. (The "Adversary," or Satan of the Old Testament). People in very recent times have reported abductions by beings answering to their description. There have also been face to face encounters. The biblical patriarchs, and prophets of ancient times, recorded visits from them, and described their physical appearance. "His body was like beryl, his face had the appearance of lightning, his eyes were like flaming torches,"...(Dan 10:5-6), and "His head and His hair were white like wool, like snow; and His eyes were like a flame of fire;" (Revelation 1:14). It could be that both sides of this epic battle will be manipulating mankind toward the end, each for their own purposes.

Christ Will Return Between 2010 AD and 2060 AD

Since planet Nibiru was last here in 1540 B.C., it should return by 2060 A.D. However, Christ did say that if those days were not cut short, no flesh would be saved. This means he will return before 2060. Probably after 2010 A.D., for this begins the last Sabbatical/Jubilee cycle. Sometime between 2010 and 2060, Christ will most assuredly return to earth to put down the Adversarial faction, and to rescue his followers. Humanity will then be mass evacuated to planet Nibiru in the New Jerusalem spacecraft.

Timeless Voyager Press

FOOTNOTES

Please note that the book "Unusual Mars Surface Features" used is the fourth edition. The version of "The 12th Planet" used is the hardback by Stein and Day. The book "The Monuments of Mars, A City on the Edge of Forever" used is the second edition.

[1] Sitchin, Zecharia, Genesis Revisited, pages 230-269. Hoagland, Richard, The Monuments of Mars, see special photographic section. Di Pietro, Vincent, Molenaar, Gregory, Brandenburg, John, Unusual Mars Surface Features, entire book dedicated to these anomalies.

[2] Sitchin, Zecharia, Genesis Revisited, pages 230-269. Di Pietro, Vincent, Molenaar, Gregory, Brandenburg, John, Unusual Mars Surface Features, entire book dedicated to this subject. Hoagland, Richard , Monuments of Mars, entire book dedicated to subject.

[3] Hoagland, Richard, The Monuments of Mars, page 4.

[4] Foster, Benjamin, Before the Muses, Volume I, The Babylonian "Epic of Creation," tablet VII, page 399.

[5] Sitchin, Zecharia, The 12th Planet, page 161-186, 218-224. (See drawings on page 219).

[6] Sitchin, Zecharia, The 12th Planet, pages 276-279. Sitchin, Zecharia, The Stairway to Heaven, pages 283-308. Sitchin, Zecharia, The Wars of gods and Men, pages 202-228. Hoagland, Richard, The Monuments of Mars, pages 363, 293,295,298 & 299.

[7] Sitchin, Zecharia, Genesis Revisited, pages 230-269.

[8] Hoagland, Richard, The Monuments of Mars, pages 125-127. Hoagland speculates on ages of monuments.

[9] Sitchin, Zecharia, The 12th Planet, page 160. Schellhorn, G. Cope, Extraterrestrials in Biblical Prophesy! Page 55-75. Baigent, Michael, From the Omens of Babylon, Astrology and Ancient Mesopotamia, see pages 98-99 where author agrees with me that the origin of the Hebrew religion was Mesopotamia, and the God of the Hebrews was the Mesopotamian moon deity Nannar-Sin.

[10] Foster, Benjamin, Before the Muses, Volume I, An Anthology of Akkadian

Literature, "Epic of Creation," pages 351-402.

[11] Foster, Benjamin, Before the Muses, Volume I, An Anthology of Akkadian Literature, "Atrsa-Hasis," Old Babylonian version, pages 158-167.

[12] Sitchin, Zecharia, The 12th Planet, page 164.

[13] Sitchin, Zecharia, The 12th Planet, pages 225-226.

[14] Sitchin, Zecahria, The 12th Planet, pages 220, 162.

[15] Sitchin, Zecharia, The 12th Planet, pages 222-227. Sitchin, Zecharia, Genesis Revisited, pages 230-271.

[16] Sitchin, Zecharia, Genesis Revisited, pages 157-202. Schellhorn, G. Cope, Extraterrestrials in Biblical Prophesy, 205-237. Pye, Lloyd, Everything You Know is Wrong, Book One: Human Origins, entire book dedicated to the origin of man, and other animals n earth.

[17] Baigent, Michael, From the Omens of Babylon, Astrology and Ancient Mesopotamia, pages 98-99. See also my book, The Holy Bible is An Extraterrestrial Transmission, the chapter titled: The Number of God's name.

[18] Sitchin, Zecharia, The twelfth Planet, pages 56-86.

[19] Sitchin, Zecharia, Genesis Revisited, pages 268-269.

[20] Sitchin, Zecharia, Genesis Revisited, pages 230-269.

[21] Drosnin, Michael, The Bible Code, page 94-95.

[22] Marrs, Jim, The Alien Agenda, pages 80-85. Reader's Digest, Mysteries of the Unexplained, pages 215-216. Sitchin, Zecharia, Genesis Revisited, page 284.

[23] Sitchin, Zecharia, Genesis Revisited, page 250.

[24] Sitchin, Zecharia, Genesis Revisited, pages 248-250. Hoagland, Richard, The Monuments of Mars, A City on the Edge of Forever, pages 4-5.

[25] Hoagland, Richard, The Monuments of Mars, A City on the Edge of Forever, pages 4-5.

[26] Hoagland, Richard, The Monuments of Mars, A City on the Edge of Forever, page 5. DiPietro, Vincent, Molenaar, Gregory, Brandenburg, John, Unusual Mars Surface Features, page 13 (fourth edition).

[27] Hoagland, Richard, The Monuments of Mars, A City on the Edge of Forever, pages 5-6. DiPietro, Vincent, Molenaar, Gregory, Brandenburg, John, page 13.

[28] Hoagland, Richard, The Monuments of Mars, A City on the Edge of Forever, pages 6-7. DiPietero, Vincent, Molenaar, Gregory, Brandenburg, John, Unusual Mars Surface Features, pages 15-29.

[29] Hoagland, Richard, The Monuments of Mars, page 7. DiPietro, Vincent, Molenaar, Gregory, Brandenburg, John, Unusual Mars Surface Features, pages 56 & 88.

[30] Sitchin, Zecharia, Genesis Revisited, pages 230-269. Hoagland, Richard, The Monuments of Mars, A City on the Edge of Forever, the entire book discusses these ruins in various sites on Mars, and possible links to earthly cultures. DiPitero, Vincent, Molneaar, Gregory, Brandenburg, John, Unusual Mars Surface Features, entire book dedicated to an examination of the various ruins, located over surface of Mars.

[31] Spinrad, Hyron, The World Book Encyclopedia, page 181, (distance of Mars to earth at its closest approach).

[32] Roaf, Michael, Cultural Atlas of Mesopotamia and the Ancient Near East, pages 124-125. Hoagland, Richard, The Monuments of Mars, A City on the Edge of Forever, page 352. Sitchin, Zecharia, Genesis Revisited, pages 212-218.

[33] Sitchin, Zecharia, Genesis Revisited, page 216.

[34] Sitchin, Zecharia, The 12th Planet, pages 172-173.

[35] Sitchin, Zecharia, Genesis Revisited, page 217.

[36] DiPietero, Vincent, Molenaar, Gregory, Brandenburg, John, Unusual Mars Surface Features, page 56, 98-113. Sitchin, Zecharia, Genesis Revisited, pages 230-269.

[37] Hoagland, Richard, The Monuments of Mars, A City on the Edge of Forever, pages 25-26.

[38] Hoagland, Richard, The Monuments of Mars, A City on the Edge of Forever, page 71.

[39] Hoagland, Richard, The Monuments of Mars, A City on the Edge of Forever, page 71.

[40] Sitchin, Zecharia, Genesis Revisited, pages 258-261.

[41] DiPietro, Vincent, Molenaar, Gregory, Brandenburg, John, Unusual Mars Surface Features, page 73.

[42] DiPietero, Vincent, Molenaar, Gregory, Brandenburg, John, Unusual Mars Surface Features, pages 72-73.

[43] Hoagland, Richard, The Monuments of Mars, page 326.

[44] Hoagland, Richard, The Monuments of Mars, pages 74-76.

[45] Chatelain, Maurice, Ancient Skies, Volume 18, number 6, A Martian Stonehenge?, page 3.

[46] Hoagland, The Monuments of Mars, pages 326-327, photo plates 14-17.

[47] Hoagland, Richard, pages 52-64.

[48] Hoagland, The Monuments of Mars, page 78.

[49] Hoagland, Richard, The Monuments of Mars, pages 70-71.

[50] Sitchin, Zecharia, The Stairway to Heaven, pages 299, & 302. Davis, Howard, The World Book Encyclopedia, Sphinx, page 610.

[51] Dipietro, Vincent, Molenaar, Gregory, Brandenburg, John, Unusual Mars Surface Features, page 38, see satellite altitude 35A72.

[52] Sitchin, Zecharia, Genesis Revisited, page 245.

[53] Hoagland's Mars: Volume II, The U.N. Briefing, The Terrestrial Connection, (extended version video), Richard C. Hoagland, courtesy of the Mars Mission, 1992.

[54] Hoagland, Richard, The Monuments of Mars, page 72.

[55] Hoagland, Richard, The Monuments of Mars, pages 81-82. Sitchin, Zecharia, Genesis Revisited, pages 259-260.

[56] Hoagland, Richard, The Monuments of Mars, A City on the Edge of Forever, pages 74-76.

[57] Hoagland, Richard, The Monuments of Mars, photographic plate 45. See also NASA photo frame #070A13.

Timeless Voyager Press

[58] Hoagland, Richard, The Monuments of Mars, see photo plate 11, in special photographic section, orthographically-rectified photo mosaic, composed of a series of computer enhanced original Viking frames (35A71,72,73,74) on opposite page. See also Hoagland's Mars: They NASA Cydonia Briefings, Volume I (video).

[59] The Crop Circle Enigma, edited by Ralph Noyes, page 172. The Atlas of Mysterious Places, edited by Jennifer Westwood, page 34, side margin, page 35, see photo. Sitchin, Zecahria, When Time Began, pages 35-36, & 69-70. Hoagland's Mars, The NASA Cydonia Briefings,: Volume I, (video). Hoagland Richard, The Monuments of Mars, pages 269,275, & 276.

[60] Hoagland, Richard, The Monuments of Mars, see text of photo plate 17.

[61] Sitchin, Zecharia, The Wars of Gods and Men, page 153.

[62] Sitchin, Zecahria, The 12th Planet, page 370.

[63] Sitchin, Zecahria, The 12th Planet, pages 369-370.

[64] Hoagland, Richard, The Monuments of Mars, page 88. Hoagland's Mars, The NASA Cydonia Briefings, vol.1, 1991, Haogland and Curley.

[65] Hoagland, Richard, The Monuments of Mars, page 88.

[66] Hoagland, Richard, The Monuments of Mars, pages 88-89.

[67] Wilkinson, Richard, Reading Egyptian Art, page 131.

[68] Hoagland, Richard, The Monuments of Mars, pages 266-268 (Egypt's Connection with Mars).

[69] Hoagland, The Monuments of Mars, page 165. Hoagland's Mars, The NASA-Cydonia Briefings, Volume I (video).

[70] Sitchin, Zecharia, The 12th Planet, pages 305-306.

[71] Hoagland, Richard, The Monuments of Mars, page 363, phorographic plate 39.

[72] Hoagland, Richard, The Monuments of Mars, page 363, photographic plate number 39.

[73] Sitchin, Zecharia, Genesis Revisited, page 202-203 (Enki's cult symbol of the serpent).

[74] Hoagland, Richard, The Monuments of Mars, page 289. DiPietro, Vincent, Molenaar, Gregory, Brandenburg, John, Unusual Mars Surface Features, page 44.

[75] Sitchin, Zecahria, Genesis Revisited, page 256. Sitchin, Zecharia, The Stairway to Heaven, pages 302-303.

[76] Sitchin, Zecharia, The Stairway to Heaven, page 113. Sitchin, Zecahria, The Wars of Gods and Men, page 128.

[77] Sitchin, Zecharia, Genesis Revisited, pages 256-257. Sitchin, Zecharia, The Stairway to Heaven, page 305.

[78] Sitchin, Zecahria, Genesis Revisited, page 257. Sitchin, Zecahria, The Stairway to Heaven, page 302.

[79] Sitchin, Zecahria, Genesis Revisited, page 257.

[80] Sitchin, Zecahria, The Stairway to Heaven, page 302. Sitchin, Zecharia, Genesis Revisited, page 257.

[81] Sitchin, Zecharia, The Stairway to Heaven, page 302.

[82] Hoagland, Richard, The Monuments of Mars, page 363.

[83] Hoagland, Richard, The Monuments of Mars, pages 362-363.

[84] Sitchin, Zecharia, Genesis Revisited, pages 256-257.

[85] Hoagland, Richard, The Monuments of Mars, page 363.

[86] Sitchin, Zecahria, Genesis Revisited, page 257.

[87] Sitchin, Zecharia, The 12th Planet, page 163.

[88] Sitchin, Zecharia, Genesis Revisited, page 257.

[89] Sitchin, Zecharia, The 12th Planet, page 100.

[90] Hoagland, Richard, The Monuments of Mars, frame #12 shows photo of teterahedral pyramid and cliff face.

[91] Steckling.Fred, We Discovered Alien Bases on the Moon!, pages 129 (NASA photo plate #16-19238, page 128, LO III 162-M, page 135 (11-37-5438).

[92] Chatelain, Maurice, Ancient Skies, Volume I, 18, Number 6, A Martian Stonehenge?, page 3.

[93] Chatelain, Maurice, Ancient Skies, Volume I, 18, Number 6, A Martian Stonehenge? Page 3-4.

[94] Chatelain, Maurice, Ancient Skies, Volume I, 18, number 6, A Martian Stonehenge? Page 4.

[95] Sitchin, Zecahria, Genesis Revisited, pages 268-269.

[96] Sitchin, Zecahria, Genesis Revisited, pages 267.

[97] The New American Standard Bible, The Lockman Foundation, side margin, verses 16-17.

[98] DiPietero, Vincent, Molenaar, Gregory, Brandenburg, John, Unusual Mars Surface Features, pages 98-103.

[99] DiPietro, Vincent, Molenaar, Gregory, Brandenburg, John, Unusual Mars Surface Features, page 101.

[100] See NASA Viking photo 86A10.

[101] See NASA Viking photo 86A10.

[102] Sitchin, Zecharia, Genesis Revisited, page 261.

[103] Sitchin, Zecharia, Genesis Revisited, page 280.

[104] Sitchin, Zecaharia, Genesis Revisited, page 280. DiPietro, Vincent, Molenaar, Gregory, Brandenburg, John, page 93.

[105] DiPietro, Vincent, Molenaar, Gregory, Brandenburg, John, Unusual Mars Surface Features,, page 93. Sitchin, Zecharia, Genesis Revisited, page 242-243.

[106] Hoagland, Richard, The Monuments of Mars, pages 320-321.

[107] Sitchin, Zecharia, The Stairway to Heaven, pages 181-182, 231-234, & 293. Sitchin, Zecharia, Genesis Revisited, pages 242-243.

[108] Hoagland, Richard, The Monuments of Mars, pages 320-321.

[109] Hoagland, Richard, The Monuments of Mars, pages 320-321.

[110] Hoagland, Richard, The Monuments of Mars, pages 320-321.

[111] Sitchin, Zecharia, Genesis Revisited, page 302.

[112] Hoagland, Richard, The Monuments of Mars, page 321.

[113] Coe, Michael, D., Breaking the Maya Code, page 27.

[114] Sitchin, Zecahria, Genesis Revisited, page 246. DiPietro, Vincent, Molenaar,

Gregory, Brandenburg, John, Unusual Mars Surface Features, pages 46-47, 97, & 45. Hoagland, Richard, Monuments of Mars, pages 9,193,206, & 207.

[115] Sitchin, Zecharia, Genesis Revisited, pages 53-54. DiPietro, Vincent, Molenaar, Gregory, Brandenburg, John, Unusual Mars Surface Features, page 57.

[116] DiPietro, Vincent, Brandenburg, John, Molenaar, Gregory, Unusual Mars Surface Features, pages 57-76.

[117] Hoagland, Richard, The Monuments of Mars, page 148-149.

[118] Hoagland, Richard, The Monuments of Mars, pages 148-149.

[119] Cohen, Michael, The war of the Gods, UFO Annual 1977, pages 36-39, 72-73.

[120] Sitchin, Zecharia, Genesis Revisited, pages 272-273.

[121] Sitchin, Zecharia, Genesis Revisited, page 274.

[122] Sitchin, Zecahria, Genesis Revisted, page 275.

[123] Sitchin, Zecharia Genesis Revisited, pages 281-283.

[124] Sitchin, Zecharia, Genesis Revisited, pages 283-284.

[125] Steckling, Fred, We Discovered Alien Bases on the Moon!, pages 129,128, 130, 135.

[126] Sitchin, Zecharia, Genesis Revisited, page 280.

[127] Ft. Worth Star Telegram, "Observer Hangs Heavy on Scientists" (August 25th, 1993), Associated Press.

[128] UFO Magazine, Volume 8, no. 6, "Did NASA Disobey the prime directive?" Richard C. Hoagland, pages 40-42, August 25, 1993.

[129] Cohen, Michael, Saga's Flying Saucer Special, UFO Annual 1977, "The War of the Gods," page 39.

[130] Cohen, Michael, Saga's Flying Saucer Special, UFO Annual 1977, "The War of the Gods," page 39.

[131] Cohen, Michael, Saga's Flying Saucer Special, UFO Annual 1977, "The War of the Gods," page 39. Frank Edwards, Flying Saucers, Serious Business, page 199-202.

[132] Barnstone, Willis, The Other Bible, page 496. Laurence, Richard, (translator) The Book of Enoch the Prophet, page 68.

[133] Cohen, Michael, Saga's Flying Saucer Special UFO Annual 1977, "The War of the Gods," page 39.

[134] Steckling, Fred, We Discovered Alien Bases On the Moon, pages 16-18 (see photo plate number 4).

[135] Sitchin, Zecharia, The 12th Planet, page 162-163.

[136] Steckling, Fred, We Discovered Alien Bases on the Moon, page 11.

[137] Steckling, Fred, We Discovered Alien Bases on the Moon, page 120. See NASA photo plate number IAU-308.

[138] Steckling, Fred, We Discovered Alien Bases on the Moon, pages 135,128-129. See NASA photographic plates #: Apollo 11, 11-37-5438, LO III 162-M, and Apollo 16-19238.

[139] Steckling, Fred, We Discovered Alien Bases on the Moon, pages 12-14, See photo plate 2. Childress, David, Hatcher, Extraterrestrial Archaeology, page 231.

Timeless Voyager Press

[140] Steckling, Fred, We Discovered Alien Bases On The Moon, pages 130-132, 137, and 138. See NASA photo frame numbers: As 13-60-8622, AS 13-60-8609, 13-60-8609, and AS-12-497319.

[141] Steckling Fred, We Discovered Alien Bases on the Moon! Page 146, NASA frame #AS-12-497319.

[142] Steckling, Fred, We Discovered Alien Bases on the Moon! Page 138, NASA frame #AS 12-497319.

[143] Wilson, Don, Our Mysterious Spaceship Moon, page 136.

[144] Wislon, Don, Our Mysterious Spaceship Moon, pages 136-137.

[145] Wilson, Don, Our Mysterious Spaceship Moon, page 138.

[146] Wilson, Don, Our Mysterious Spaceship Moon, pages 138-139.

[147] Wilson, Don, Our Mysterious Spaceship Moon, page 139.

[148] Wilson, Don, Our Mysterious Spaceship Moon, page 139-140.

[149] Wilson, Don, Our Mysterious Spaceship Moon, page 140-141.

[150] Wilson, Don, Our Mysterious Spaceship Moon, page 141-142.

[151] Wilson, Don, Our Mysterious Spaceship Moon, page 60. Shoemaker, Eugene, World Book Encyclopedia, page 646f, "Moon."

[152] Wilson, Don, Our Mysterious Spaceship Moon, pages 142-143.

[153] Wilson, Don, Our Mysterious Spaceship Moon, page 143.

[154] Wilson, Don, Our Mysterious Spaceship Moon, page 144-145.

[155] Steckling, Fred, We Discovered Alien Bases on the Moon, page 40, see drawing.

[156] Wilson, Don, Our Mysterious Spaceship Moon, page 145.

[157] Wilson, Don, Our Mysterious Spaceship Moon, page 54-55.

[158] Wilson, Don, Our Mysterious Spaceship Moon, pages 47-48.

[159] Watson, Celestial Raise, Richard Watson, and ASSK, pages 147-148.

[160] Chatelain, Maurice, Our Cosmic Ancestors, page 24.

[161] Steckling, Fred, We Discovered Alien Bases On the Moon!, pages 8, 57-58, see photo plates 42-45.

[162] The World Book Encyclopedia, "Moon" page page 646, Shoemaker, Eugene M.

[163] Steckling, Fred, We Discovered Alien Bases On the Moon!, pages 61-63, 75, 83, 84, and 86. See NASA photo frames: HR 128, 151 (3) MR121 (1)-2.

[164] Steckling, Fred, We Discovered Alien Bases on the Moon!, pages 66-67.

[165] Steckling, Fred, We Discovered Alien Bases on the Moon! Pages 73, and 92-93 (see photos).

[166] Steckling, Fred, We Discovered Alien Bases on the Moon!, page 66.

[167] Chatelain, Maurice, Our Cosmic Ancestors, page 25.

[168] Chatelain, Maurice, Our Cosmic Ancestors, page 25. Durfield, Yvonne S., Ideal's UFO Magazine, number 4, 1978, Astronaut's Strange Fates, Linked to UFO's, page 46.

[169] Chatelain, Maurice, Our Cosmic Ancestors, page 25.

[170] Edwards, Frank, Flying Saucers, Serious Business, page 317.

BIBLIOGRAPHY

[1] Roaf, Michael, Cultural Atlas of Mesopotamia and the Ancient Near East, Equinox Ltd., Musterlin House, Jordan Hill Rd., Oxford, OX2 8DP, England, copyright 1990.

[2] Reader's Digest, Mysteries of the Unexplained, The Reader's Digest Association, Inc., Pleasantville, New York, Montreal, copyright 1982.Sitchin, Zecharia, The 12th Planet, Stein and Day, Scarborough House, Briarcliff Manor, New York, 10510, copyright 1976.

[3] Sitchin, Zecharia, Genesis Revisited, Avon Books, A Division of the Hearst Corporation, 1350 Avenue of the Americas, New York, New York, 10019, copyright 1990.

[4] Marrs, Jim, The Alien Agenda, Harper Collins Publishers, Inc., 10 east 53rd street, New York, N.Y. 10022, copyright 1997.

[5] Hoagland, Richard, The Monuments of Mars, A City on the Edge of Forever, North Atlantic Books, 2800 Woolsey St., Berkeley, California 94705, copyright 1987, 1992.

[6] Wilson, Don, Our Mysterious Spaceship Moon, Dell publishing company Inc., 1 Dag Hammarskjold Plaza, New York, N.Y., 10017, copyright 1975.

[7] Sitchin, Zecharia, The 12th Planet, Stein and Day publishers, Scarborough House, Briarcliff Manor, N.Y. 10510, copyright 1976.

[8] Chatelain, Maurice, Our Cosmic Ancestors, Temple Golden Publications, P.O. Box 10501, Sedona, Arizona, 86336, copyright 1987.

[9] Wilkinson, Richard H., Reading Egyptian Art, A Hieroglyphic Guide to Ancient Egyptian Painting and Sculpture, Thames and Hudson Inc., 500 fifth

Avenue, New York, New York, 10110, copyright 1992.

[10] Baigent, Michael, From the Omens of Babylon, Astrology and Ancient Mesopotamia, Penguin Books, Ltd., 27 Wrights Lane, London W85TZ, England, copyright 1994.

[11] Steckling, Fred, We Discovered Alien Bases on the Moon, Fred Steckling, P.O. Box 1722, Vista California 92805, copyright 1981.

[12] Shoemaker, Eugene, Spinrad Hyron, The World Book Encyclopedia, Volume M, "Moon," "Mars," Field Enterprises Educational Corporation, copyright 1971.

[13] DiPietro, Vincent, Molenaar, Gregory, Brandenburg, John, Unusual Mars Surface Features, fourth edition, Mars Research, P.O. Box 284, Glenn Dale, Maryland, 20769.

[14] Chatelain, Maurice, Ancient Skies, Volume 18, number 6, A Martian Stonehenge?, page 3, 1992.

[15] Sitchin, Zecharia, The Stairway to Heaven, Avon Books, A Division of the Hearst Corporation, 105 Madison Avenue, New York, N.Y. 10016, copyright 1980.

[16] Hoagland's Mars: Volume II, The U.N. Briefing, The Terrestrial Connection (extended version video), courtesy of the Mars Mission 1992.

[17] Hoagland's Mars: The NASA Cydonia Briefing, Volume I, (video).

[18] The Crop Circle Enigma, A Range of Viewpoints from the Center of Crop Circle Studies, edited by Ralph Noyes, Gateway Books, The Hollies Wellow, Bath BA2, 82J, UK, copyright 1990.

[19] Sitchin, Zecharia, When Time Began, A Division of the Hearst Corporation, 1350 Avenue of the Americas, New York, N.Y. 10019, copyright 1993.

[20] Westwood, Jennifer, The Atlas of Mysterious Places, Marshall Editions Ltd., & Weidenfield & Nicholson, New York, A Division of Wheatland Corporation, 10 East 53rd Street, New York, N.Y. 10022, copyright 1987.

[21] Sitchin, Zecharia, The Wars of Gods and Men, Avon Books, A Division of the Hearst Corporation, 105 Madison Avenue, New York, N.Y. 10016, copyright 1985.

[22] The Other Bible, edited with introductions by Willis Barnstone, copyright 1984.

[23] Coe, Michael, Breaking the Maya Code, 1992.

[24] Ft. Worth Star Telegram, "Observer Hangs Heavy on Scientists" (August 25, 1993), Associated Press.

[25] Hoagland, Richard, UFO (magazine), Volume 8, no. 6, "Did NASA Disobey the Prime Directive?" 1993.

[26] Edwards, Frank, Flying Saucers, Serious Business, Citadel Press, 120

Enterprise Avenue, Secausus, N.J., 07094, copyright 1966.

[27] Durfield, Yvonne S., Ideal's UFO Magazine number 4, 1978, "Astronaut's Strange Fates Linked to UFO's.

[28] Watson, Richard, & ASSK, Celestial Raise, 1987.

[29] Childress, Hatcher, David, Extraterrestrial Archaeology, Adventures Unlimited Press, Stelle, Illinois 60919, copyright 1994.

[30] Cohen, Michael, Saga's Flying Saucer Special UFO ANNUAL 1977, "The War of the Gods."

Timeless Voyager Press

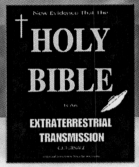
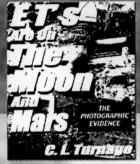

Cassette tapes of Timeless Voyager Radio Programs

108-T,233-T,234-T CASEY, Joel: Spirit Guides($9.95+$2 Shipping or all 3 $24.95+$4 shipping)
If you've ever wanted an understanding about the nature of spirit guides, these are the tapes for you.

242-T, 249-T CLEMENTE, Kasandra: St. Germaine ($9.95+$2 Shipping or both for $16.95+$3 shipping)
Channeled entity, St. Germaine answers questions from Bruce and listeners. Very provacative interview.

109-T ESSENE, Virgina: New Teachings / The Christ Energy ($9.95+$2 Shipping)
Author, Virginia Essene, "New Teachings", "New Cells / New Bodies", and other books talks very easily about embedded genetic codes which are placed in our body format.

110-T FICKES, Bob: Ascension($9.95+$2 Shipping)
Renowned Spiritual Teacher, Shaman, Author, and Channel for the Ascended Masters, Bob Fickes explains 4th and 5th dimensionality.

111-T FOWLER, Chad: Clear Channel Counseling($9.95+$2 Shipping)
Clear Channeling is quite a bit different from trance-channeling in that the channel is stepping aside and receiving the information while still aware.

113-T JOHANNSEN, Susan: "Philip"($9.95+$2 Shipping)
Susan is the publisher of a newsletter, "HOMEWORDS", which is filled with information channeled from the entity, "Philip". During the interview Bruce asks Philip some very pertinent questions concerning the nature and meaning of dreams and the veil of ignorance.

114-T KATAR: "Tseng Tsing" / "Lord Michael", Channeled Entities($9.95+$2 Shipping)
What is the difference between whole body channeling and channeling? Why bring Spirit entities in thru your body as a vehicle? These questions and many more are answered by Katar. She channels the Master, Tseng Tsing and Archangel, Michael during this interview.

115-T, 116-T KATAR: "Sanat Kumara" / "Clark" / "Mother Mary" / "Jesus" Pt 1&Pt 2($16.95+$3 shipping)
Katar channels for approximately 2 hours. Bruce has a very interesting discourse with CLARK about many subjects including "unconditional love". This is a very special two tape set for the true seeker of great knowledge.

117-T MILLS, Gloria: Channel / Ascended Masters / Hierarchy ($9.95+$2 Shipping)
Gloria Mills, explores many areas of interest including: The hierarchy of entities, their names, the types of information they wish to bring forth, spirit guides, and a lot more.

118-T NEEB, Patricia: Channel($9.95+$2 Shipping)
Clinical Psychologist, Patricia Neeb uses channeling in her practice and also teaches breathing and grounding-type techniques.

226-T OTERO, Steve: The Deep Meditation Experience($9.95+$2 Shipping)
Steven Otero explains in great detail what meditation is, and gives the listener two different short guided meditation adventures.

Order tapes @
1-800-576-8463

Timeless Voyager Press

Cassette tapes of Timeless Voyager Radio Programs

224-T DORLAND, Frank: Effects of Crystals on the Human($9.95+$2 Shipping)
Biocrystallography-the study of the interchange of energies between the human mind and the electronic quartz crystal. The crystal responds to what it is told to do. He explains: Programming a crystal, cutting it correctly, silver vs. gold ornamentation, amplification, and much more. The Best!

136-T CHOPRA, Dr. Deepak: Ayurvedic Medicine($9.95+$2 Shipping)
In this interview and address, Dr. Chopra discusses Ayurvedic medicine as the tool to intervene at the level where we are being created new each day.

250-T CHOPRA-DOUILLARD:Ayurvedic Medicine/Sports($9.95+$2 Shipping)
This is a very important interview with both Dr Deepak Chopra and Dr. John Douillard. While many are familiar with Chopra's work in the health fields, many are not yet familiar with the application of this ancient medicine for training the human body. Very fresh info covers Chopra's new work and Douillards sports successes.

144-T McCABE, Ed: Oxygen Therapies($9.95+$2 Shipping)
For over 100 years these little known therapies have been used to cure such diseases as Lukemia, Diabetes, Allergies, and many more. Now recent investigation by journalist, Ed McCabe is showing that Cancer, Herpes, Flu, and even Aids can also be cured.

251-T ANDERSON, Joan Wester: Where Angels Walk($9.95+$2 shipping)
Author, Joan Anderson has compiled a book with about 60 short stories of Angel encounters by people from all walks of life. She even starts out by telling us about her son's encounter and near-fatal evening when saved by an impossible set of circumstances.

230-T ANDERSON, Richard Feather: Geomancy($9.95+$2 shipping)
Geomancy is ancient art of finding the right place for almost anything. Place to sit, sleep, build a house, look for water, and etc. Richard explains and teaches this art.

214-T ANILDA, Rev. Elizabeth: The 12 Rays($9.95+$2 Shipping)
If you've ever wondered about where all of the planes of reality are, or perhaps how they would stack up moving from the earth plane outwards, here is the interview for you. Rev. Elizabeth Anilda explains where the Ascended Masters are in the scheme of this hierarchy, the differences between the Cherubim and Seraphim, the Elohim, the secret rays, and more !

242-T BROWN, Rick: "Think and Grow Breasts"($9.95+$2 Shipping)
It sounds like a joke, right? It's actually a verifiable way for women who are concerned about their breast size to do something about it. Listener questions are very pertinent. While the subject matter sounds strange for a Timeless Voyager interview, it becomes evident that through hypnosis we can change anything!

225-T CLARK, Iris: Keys Of Enoch($9.95+$2 Shipping)
97 years "young", Iris Clark teaches classes in Sedona Arizona on the KEYS OF ENOCH by Dr. J.J. Hurtak. The book is one of the most sought after scientific examinations of the up-coming changes in 3

Order Tapes @
1-800-576-8463

Cassette tapes of Timeless Voyager Radio Programs

149-T,273-T COOK, Bob: Cook Inertial Propulsion Engine($9.95+$2 or both for $16.95+$# shipping
(149-T) Inventor, Bob Cook explains his amazing propulsion system which uses no fossil fuels. MIT engineer, Dick Rose and JPL engineer, John Lu Valle both corroborate the info via conference call hook-up in this incredible interview. The system is the first motion causing machine to defy the third law of motion as established by Sir Isaac Newton, the father of physics.(273-T) WWCR shortwave with listener call-ins from around the world.

150-T, 151-T CREME, Benjamin: The Reappearance of the Christ($9.95+$2 Shipping or both for $16.95+$# shipping)
Author and lecturer, Benjamin Creme gives a very clear and concise understanding of who Jesus is and where he is now. He also answers some very important questions regarding Lord Maireya, the embodiment of the Christ energy, which world leaders have actually seen him, the 5 step Initiation which we are all going thru, and what we can do to be part of the new awakening.

152-T CURTIS, Chara: All I See Is Part Of Me($9.95+$2 Shipping)
Thru the eyes of a child in this story of a mystical journey, we begin to discover our own common link with the totality of life.

153-T GURNEY, Carol: Animal Communication($9.95+$2 Shipping)
Carol Gurney communicates with animals telepathically. During the interview, Bruce asks many questions which give the listener an in-depth understanding of how animals communicate.

223-T HOLMS, Bruce Stephen: LIVE on KCSB($9.95 +$2 Shipping)
Rare TVR show with no guest...just Host and Producer, Holms talking about UFO's and Earthquake predictions along with listener calls.

156-T, 220-T MAXWELL, Jordan: Religeo-Political Philosophy ($9.95+$2 Shipping or both for $16.95+$3 shipping)
A very controversial show regarding the beginnings of Christianity. He also speaks about the manipulative control that the Vatican has on the politics of all nations, the Illuminati, and the Free Masons who actually control all governments.

241-T MAXWELL, Jordan: Religeo-Political Philosophy ($9.95+$2 Shipping
Still controversial, Jordan is live on TVR's flagship station and not only gives us more info about symbology etc., but also answers listeners questions.

221-T,228-T,274-T ROBERTS, Rozella: UFO's, Spirit Voices, etc. ($9.95+$2 S/H, 2 for $16.95+$3 S/H, 3 for $24.95+$4 S/H)
(221-T) "Matriarch of Unexplained Phenomena", 82 year old Rozella Roberts, author, book publisher, speaker, UFO enthusiast, and Spirit Voice/Spirit Photo investigator takes the listener on a fast paced tour of her many years of exploration. (228-T) In this interview Rozella Roberts and her long time researcher/companion Evelyn Chiaverini discuss the "Electronic Voice Phenomena". EVP as it is known today is a unique method that anyone can use to speak to the dead, dis-embodied, extra-terrestrials or even plants and animals. (274-T) Listener call-ins from around the world + Ashtaar message.

Order Tapes @
1-800-576-8463

Cassette tapes of Timeless Voyager Radio Programs

181-T ANDREWS, Colin: Crop Circles Phenomenon ($9.95+$2 Shipping)
Colin Andrews is the leading expert on the CROP CIRCLE phenomenon. The interview is filled with incredible information including a description of a video of a UFO going through the fields... flying around these "Agri-glyphs".

184-T BEDELL, Barry: Urantia#1 ($9.95+$2 Shipping)
This is the book that covers in 2,097 pages, the history of the Universes, the earth (which by the way is called "URANTIA"), and the Life and Teachings of Jesus.

185-T BELL, Dr. Fred: UFO/ Pyramid/Power($9.95+$2.Shipping)
The Powerdome, Nuclear Receptor, Nuclear Disc, Pyradyne Laser, The Seven Sub-planes of the Physical,Geometric Force Field Patterns, The Orb - Irradiator Relationship, and The Omnion. Dr. Bell tells Bruce what these devices are used for and then goes into a very in depth dialogue on the "GREYS" (negative extra-terrestrials).

186-T, 187-T BIELEK, Alfred: Philadelphia Experiment ($9.95 +$2 shipping or both for $16.95+$3 shipping)
Al Bielek is a survivor of the infamous Navy invisibility experiment which caused a large battle ship to become optically and radar invisible on August 12, 1943. His experience of interdimensional time travel into the future of 1983 and back in a matter of hours is incredible.

238-T CANADA, Steve: Crop Circles Explained($9.95+$2 Shipping)
Steve Canada not only explains the Crop Circle phenonmena, but adds a great deal to the "mystery" by explaining his discovery of who (ET) is making them and why. Includes very interesting factual data!

188-T CHILDRESS, David Hatcher/Vimana Aircraft($9.95+$2 Shipping)
After researching the Vimaanika Shastra by Mahrishi Bharadwaaia he discovered that there were actually Flying Saucer type vehicles over 4,000 years ago constructed from the explicit instructions contained in the manuscript.

189-T COHAN, Norman: "Sumerians/UFO's/Planet Nibiru"($9.95+$2 Shipping)
Here, is a lucid explanation on how we (humans) got here, why we were engineered, who engineered us, and why we call them GODS.

234-T, 235-T COLE, Yvonne: Ashtar Command($9.95+$2 Shipping, 2 for $16.95+$3 shipping)
Telepathic receiver for the Ashtar Command, Yvonne Cole brings through some very important information (234-T) in this channeled message. In this interview (235-T), Yvonne explains who the "GREYS" are and what they are doing in the "so-called" abductions and much more!

Order tapes @
1-800-576-8463

Cassette tapes of Timeless Voyager Radio Programs

240-T HEERMAN, David Lee: "Operation Grey Strike"($9.95+$2 Shipping)
Heerman claims to have discovered several sights in Los Angeles that have an alien presence on them. "Tingle Fields" and "Laminers" are invisible to the eye but perceptible to the touch. He believes that they insight us to high excited states in order to drain the energies from the our energy fields or Auras. Incredible evidence and special "sound" for annilating these aliens played for 45 seconds.

231-T HOAGLAND, Richard C.: "Face On Mars"($9.95+$2 Shipping)
Hoagland has discovered an ancient city complete with pyramids, a sphinx like sculpture called "The Face" in Cydonia. At NASA's request he presented hius findings to 1,000's of NASA engineers now NASA has decided to black out all pictures of the August 1993 fly by.

226-T HOAGLAND, Richard C.: "Hyjacking of Mars Mission" ($9.95+$2 Shipping)
In this incredible interview Hoagland explains why he believes that a small renegade group of space scientists have hyjacked the Mars probe and why NASA is going along covering-up with the "It doesn't work" synopsis.

154-T, 192-T HOWE, Linda Moulton: Alien Harvest *Warning: Very controversial material.* ($9.95+$2 Shipping or both for $16.95+$3 shipping)
Linda is an EMMY winning television producer who has written an intriguing book about the thousands of reported animal mutilations throughout the US and world.

193-T LAWRENCE, Robert: Urantia *Warning: Very controversial material.* ($9.95+$2 Shipping)
Robert discusses the Life and Teachings of Jesus according to the URANTIA book. Because of his theological background, the interview is very stimulating, and at times "eye-opening".

194-T, 195-T LINDEMAN, Michael: "UFO Presence/ Cover-up" ($9.95+$2 Shipping or both for $16.95+$3 shipping)
Author and speaker Michael Lindeman talks freely about the US Governments attempts to cover-up evidence of UFO's and Extra-terrestrial presence in Nevada.

243-T LINDEMAN, Michael: "UFO Presence / Cover- up" ($9.95+$2 Shipping, or 3 for $24.95+$4 shipping)
Lindeman speaks very frankly about his UFO agenda and how it differs from many "run-of-the-mill" UFOlogists. Listener questions

189-T COHAN, Norman: "Sumerians/UFO's/Planet Nibiru"($9.95+$2 Shipping)
Here, is a lucid explanation on how we (humans) got here, why we were engineered, who engineered us, and why we call them GODS.

234-T, 235-T COLE, Yvonne: Ashtar Command($9.95+$2 Shipping, 2 for $16.95+$3 shipping)
Telepathic receiver for the Ashtar Command, Yvonne Cole brings through some very important information (234-T) in this channeled message. In this interview (235-T), Yvonne explains who the "GREYS" are and what they are doing in the "so-called" abductions and much more!

Order tapes @
1-800-576-8463

Printed in the United States
110592LV00001B/247/A